Sustainability Performance Evaluation of Renewable Energy Sources: The Case of Brazil

João Fernando Gomes de Oliveira
Tatiana Costa Guimarães Trindade

Sustainability Performance Evaluation of Renewable Energy Sources: The Case of Brazil

Springer

João Fernando Gomes de Oliveira
São Carlos School of Engineering
University of São Paulo
São Carlos, São Paulo
Brazil

Tatiana Costa Guimarães Trindade
São Carlos School of Engineering
University of São Paulo
São Carlos, São Paulo
Brazil

ISBN 978-3-030-08493-6 ISBN 978-3-319-77607-1 (eBook)
https://doi.org/10.1007/978-3-319-77607-1

Printed on acid-free paper

This Springer imprint is published by the registered company Springer International Publishing AG part of Springer Nature
The registered company address is: Gewerbestrasse 11, 6330 Cham, Switzerland

Foreword

Energy in modern society is an essential ingredient for a wealth economy, like the blood circulating in our arteries.

Since the beginning of civilization, humans have learned how to use biomass combustion for cooking and heating, the wind force for navigation, and the rivers flow to move mills. With increasing world population, material progress in various parts of the world, and, especially, with the formation of large cities, new sources of energy have become necessary as coal, oil, and gas and more recently nuclear power.

In the early twentieth century, electricity production moved to the forefront as it can be converted to work with greater efficiency than coal, oil, and gas. In this century, the electrification of the world economy is intensifying significantly, since electricity consumption is increasing about 4–5% per year while the energy consumption as a whole is only growing about 2% per year. Hydroelectric, wind, and photovoltaic energy—which are renewable in contrast to electricity produced by fossil fuels—have become a priority.

In 2017, the world investment in energy production was just over US$ 1.7 trillion. US$ 718 billion was invested in the production of electricity and the remainder in the production of oil, gas, coal, and energy efficiency. Of these US$ 718 billion, US$ 277 billion was invested in power grids, US$ 193 billion in electricity generated from fossil fuels, and US$ 297 billion in renewable energy, which in 2016 were distributed as follows: wind power (39%), photovoltaics (28%), hydroelectric (26%), and nuclear (7%).

The aim of this book is to propose criteria that help public or private investors in prioritizing these investments. As an example of the application of these criteria, the authors chose and evaluated six existing plants, using three renewable energy sources (hydroelectric, wind and photovoltaic) in two regions of Brazil, South and Northeast. The studied cases were compared using several economic, environmental, and social indicators.

Three examples of scenarios representing investors interested were formulated and evaluated: primarily with focus on the financial return (scenario 1); with emphasis on social and environmental aspects (scenario 2); and investors with a balanced view on the financial, environmental, and social aspects of the venture (scenario 3).

The final conclusion is that "although considered a clean energy source, renewable sources are not free from negative impacts. In addition, an energy source may have greater or lesser sustainability relative to other sources, depending on the region, indicators, and investors view considered in the analysis."

Only a balanced analysis can decide the best alternative. For this reason, the methodology developed in this work will certainly be very useful for public or private investors in the area of renewable energy.

São Paulo, Brazil

Prof. José Goldemberg
University of São Paulo—USP,
São Paulo Research Foundation—FAPESP

Acknowledgements

The authors would like to acknowledge the contribution of Bruna Paula Sobreiro for the careful reviewing work of figures, symbols, tables, and summary.

Contents

About the Authors

João Fernando Gomes de Oliveira Prof. João Oliveira graduated (1982) and earned his Ph.D. (1988) in Mechanical Engineering—University of São Paulo, he was visiting scholar (1993–1994) and visiting professor (2016) at Mechanical Engineering—University of California at Berkeley. He is Professor of Mechanical Engineering at University of São Paulo with expertise in Innovation Management, Advanced Manufacturing, and Industrial Sustainability. From 2008 through 2012, he served as President and CEO of the State of São Paulo Institute for Technological Research—IPT, a state-owned research company with more than 1200 collaborators. In IPT, he implemented a large research investment program in order to support the research and development in cooperation with Brazilian companies on new energy projects such as solar, wind, hydroelectric, deepwater oil exploration, and others. He has published over 200 papers in journals, conferences, magazines, and newspapers and holds five patents registered at Brazil's National Institute of Industrial Property (INPI). He received the major science awards in Brazil, such as the Great-Cross of the National Order of Scientific Merit by the Presidency of Brazil; the Conrado Wessel Foundation Award for Science and Culture; the CAPES- Anísio Teixeira Award for his contribution to education and research; the Personality of Technology Award by the São Paulo Association of Engineers; and the SAP Americas Innovation Award. Professor Oliveira engaged in joint research with the industrial sector (TRW Automotive and Saint-Gobain Abrasives) and acted as a consultant of automotive industry worldwide including BMW, Hyundai, Chrysler, among others. He was co-founder and served as President and CEO of EMBRAPII-National Enterprise for Industrial Innovation and is member of the National Council of Science and Technology of the Brazilian Government and member of the board for several research institutes and funding organizations. He is also Fellow and Vice-President of the Brazilian Academy of Sciences and Fellow of the International Academy for Production Engineering (CIRP).

Tatiana Costa Guimarães Trindade Graduated in Environmental Engineering from the School of Engineering of São Carlos (EESC), University of São Paulo. Through the *Sciences Without Borders* Program, she had undergratuate studies at the University of Toronto, Canada, where she received a broad introduction to renewable energy and sustainability. During the exchange year, she attended the Sustainability Conference at the University of Toronto where she learned about the challenges involving energy storage and the political implications of energy supply and use. She also participated in the Technological Diffusion Course in Introduction to Solar Photovoltaic Energy, offered by the State University of Campinas (Unicamp). In addition, she was part of an online training program promoted by the Renewable Energy Observatory of Latin America and the Caribbean (UNIDO), in which she enrolled in different modules on solar, wind, and biogas energy. She developed her final project in the area of evaluation of sustainability indicators of the main renewable energy sources in Brazil.

Symbols and Abbreviations

AC	Alternated-current
AHP	Analytic Hierarchical Process
ANA	National Water Agency
ANEEL	National Electricity Agency
ANP	National Petroleum, Gas and Biofuels Agency
ASI	Aggregate Sustainability Index
BEN	National Energy Balance
BNDES	National Economic and Social Development Bank
BRL	Brazilian Currency (Real)
COGERH	Water Resources Management Company
CSD	Commission on Sustainable Development
DC	Direct-current
EEA	European Environmental Agency
EISD	Energy Indicators for Sustainable Development
EPE	Energy Research Company
Eurostat	Statistical Office of the European Communities
FU	Federative Unit
GDP	Gross Domestic Product
GHG	Greenhouse Gases
GRI	Global Report Initiative
GW	Gigawatt
GWEC	Global Wind Energy Council
GWh	Gigawatt-hour
HPP	Hydroelectric Power Plant
IAEA	International Atomic Energy Agency
IBGE	Brazilian Institute of Geography and Statistics
IC	Total Cost to Install the Power Generation
IEA	International Energy Agency
INPE	National Institute of Spatial Research
IP	Installed Power

IPCC	Intergovernmental Panel on Climate Change
IPEA	Institute of Applied Economic Research
ISED	Indicators for Sustainable Energy Development
JPOI	Johannesburg Plan of Implementation
kV	kilovolt
kVA	kilovolt-ampere
kW	kilowatt
kWh	kilowatt-hour
LCA	Life Cycle Analysis
LPG	Liquefied Petroleum Gas
MDG	Millennium Development Goals
MJ	Mega joules
MME	Mines and Energy Ministry
Mtoe	Million Tonnes of Oil Equivalent
MW	Megawatts
MWh	Megawatt-hour
PROINFA	Alternative Energy Sources Incentive Program
PSPP	Photovoltaic Solar Power Plant
PV	Photovoltaic Cell
RFS	Reduced Flow Section
SDG	Sustainable Development Goals
SHP	Small Hydroelectric Plant
SIN	National Interconnected Electric System
TWh	Terawatt-hours
UN	United Nations
UNDESA	United Nations Department of Economic and Social Affairs
USD	United States currency (Dollar)
WEC	World Energy Council
WPP	Wind Power Plan
WSSD	World Summit on Sustainable Development
WWEA	World Wind Energy Association
WWF	Worldwide Fund for Nature

List of Figures

List of Tables

Introduction

Energy is an indispensable element for the survival and development of modern society. From the use of heat in industries to the heating of a house, from feeding a freighter to fueling a motorcycle, from automating an entire production process to lighting an office lamp, energy use is an essential part of our everyday life across different scales and forms of utilizing it. With the development of new technologies, coupled with the exorbitant growth of the world population and the emergence of new needs, one of the great challenges we currently face is the uncertainty of energy availability in the future given its current demand, primarily due to the extensive use of nonrenewable sources and the large-scale exploitation of the planet's natural resources.

Since the advent of the Industrial Revolution, energy has become intrinsically associated with the economic competitiveness of nations and their quality of life. In this context, economies with greater access to inexpensive energy resources, with low environmental impact, guarantee significant market advantages. Therefore, the search for new renewable energy sources reflects not only the current need for diversifying the world energy matrix, but also the inevitability of developing alternatives that can reduce the impacts caused by the generation, distribution, and consumption of energy. It is from this perspective that discussions arise regarding technical, economic, environmental, and social feasibility for implementing renewable energy systems.

This discussion has reached such great importance in the international context that the access to "Clean and Affordable Energy" has become one of the 17 Sustainable Development Goals (SDG) listed in the 2030 Agenda, created by the Member States of the United Nations. The 2030 Agenda is a global result of the joint work of governments and citizens aimed at creating a new model for ending poverty, promoting prosperity and social welfare, protecting the environment and combating climate change (Organização das Nações Unidas 2017). The established SDGs, successors of the Millennium Development Goals (MDGs), cover very broad but interlinked areas, ranging from equitable access to quality education and health services to energy and environmental sustainability.

More specifically, SDG No. 7 states that countries adopting the action plan set out in the Agenda should "ensure reliable, sustainable, modern and affordable access to energy for all" (UN 2017). To achieve this objective by 2030, these countries should substantially increase the share of renewable energy in the global energy matrix and strengthen international cooperation to facilitate access to research and clean energy technologies, including renewable energy, energy efficiency, and advanced fossil fuels. In addition, they should also promote investment in energy infrastructure and cleaner energy technologies (UN 2017).

Brazil was one of the 193 UN member states to adopt the 2030 Agenda, with increasing attention being paid to the discussion and search for a cleaner energy matrix. In the national context, in the coming decades, this energy issue is presented to the country as both a challenge and an opportunity (Tolmasquim 2016). A challenge, as economic and social development will require an expressive amount of energy and, consequently, a high degree of security and energy sustainability. An opportunity, given Brazil's very interesting conditions of renewable energy resources and technology to harness the energy of its various natural resources (Tolmasquim 2016).

In view of the previously elucidated problem, this book presents a methodology for the sustainability performance assessment and its application in different renewable energy production projects, in order to identify the potentials of different regions and to evaluate the sustainability of each system. Thus, a case study of electricity power plants is presented regarding different renewable energy sources in different regions of Brazil. The analysis helps in determining which projects perform better, according to different considered scenarios. The analysis is carried out based on performance indicators obtained from a survey on the characteristics of the studied projects and from the existing literature.

Besides this introductory Preface, this book has six chapters, organized as follows:

- Chapters 1–3 contain definitions, information, and data on the world energy matrix, specific, selected renewable energy sources, and sustainability indicators;
- Chapter 4 describes the main sustainability indicators used in the sustainability analysis of renewable energy generation systems and how to select them for the application in a sustainable energy endeavor;
- Chapter 5 presents a methodology for the sustainability performance analysis of different renewable energy technologies;
- Chapter 6 shows in detail the application and discussion of the sustainability performance analysis developed for a Brazilian case study, following the proposed methodology.

References

Tolmasquim MT (2016) Energia Renovável: Hidráulica, Biomassa, Eólica, Solar, Oceânica/Mauricio Tiomno Tolmasquim (coord.). EPE: Rio de Janeiro

Organização das Nações Unidas—ONU (2015) Agenda 2030. Available at <https://nacoesunidas.org/pos2015/agenda2030/>. Accessed 5 out. 2017

United Nations (2017) Sustainable development goals. Available at <http://www.un.org/sustainabledevelopment/sustainable-development-goals/>. Accessed 4 Aug 2017

Chapter 1
World Energy Matrix

Abstract An overview of the world energy matrix is presented with some specific aspects on non-renewable energy, climate changes and challenges for the development of renewable energy sources. The chapter also shows historical data on global energy consumption by energy sources and a prospection for the next 20 years. This evaluation is followed by an analysis on the Brazilian energy matrix, consumption, electrical systems characteristics, programs and incentives.

The transition from the current global energy matrix to an economically viable, environmentally less impacting and socially more equitable future scenario makes the development of renewable energy sources increasingly prominent in global efforts for a more sustainable development. This chapter presents an overview of the global energy issue, defining which sources are considered sustainable and what are the challenges for them in achieving meaningful participation in the energy market. In addition, the chapter also presents an overview of the Brazilian energy matrix, essential for better understanding the case study that will be developed in Chap. 5, which presents an application of the sustainability performance assessment method proposed in Chap. 4.

1.1 Overview of the World Energy Matrix

Global energy demand for socioeconomic development and improvement of the quality of life of the population has increased significantly due to the technological and scientific advances of human society. Since around 1850, global consumption of fossil fuels (coal, oil and gas) has dominated the energy supply, leading to the rapid growth in greenhouse gases (GHG) emissions, and the consumption of these non-renewable sources of energy is responsible for most of the global anthropogenic emissions of these gases (Intergovernmental Panel on Climate Change (IPCC) 2012).

J. F. G. de Oliveira and T. C. G. Trindade, *Sustainability Performance Evaluation of Renewable Energy Sources: The Case of Brazil*,
https://doi.org/10.1007/978-3-319-77607-1_1

However, since the last turn of the century, the world has witnessed an overt change in the consumption of energy sources. The energy sector has changed significantly, mainly due to the unexpected increase in investments in the renewable energy market, new technological capacities and high growth rates in developing countries (World Energy Council (WEC) 2016a). Consequently, we have increasingly witnessed the rapid development of unconventional sources and technological developments for the exploration of all forms of energy. This has contributed to the fall in prices and increased the disassociation between the concepts of economic growth and GHG emissions. Therefore, most countries are seeking to achieve a greater diversification of their energy matrix, with growth in the private energy microproduction sector and development in the decentralized generation system (WEC 2016a).

As can be seen in Fig. 1.1, although non-renewable energy sources continue to account for almost 90% of the world's primary energy consumption, a downward trend in fossil fuel use over the last 15 years can be seen, mainly in relation to coal and fuel oil. In addition, the participation of renewable sources, with emphasis on solar and hydropower, begins to gain a greater space in the world energy matrix.

It should be remembered that the energy sector is a long-term investment, where it may take much longer for changes to take place. Thus, when the world primary energy consumption framework is compared over the last 15 years, the changes perceived can in fact be striking (WEC 2016a). In light of this, although the progress of the renewable energy sector still does not meet optimistic expectations, it is definitely growing rapidly, representing irrefutable significance in the world energy panorama.

Efforts to transition to a cleaner energy matrix occur mainly in the electricity generation sector. Considering that this sector is responsible for approximately 25% of global GHG emissions, these still subtle advances, as observed in Fig. 1.2, are considered to be very positive due to the high system inertia for changes (WEC 2016a).

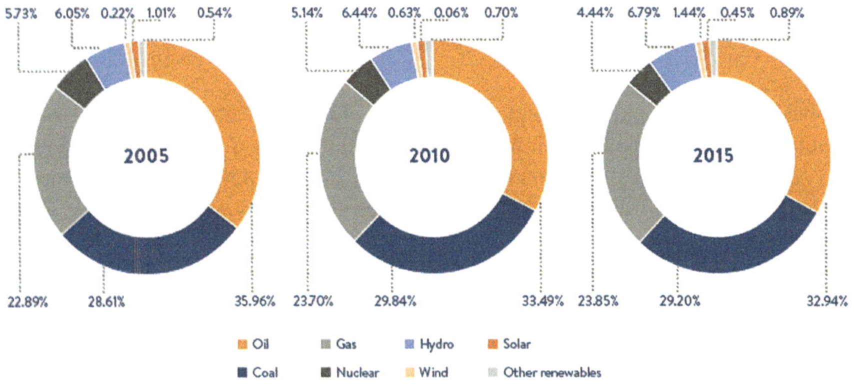

Fig. 1.1 Comparative scenarios of world primary energy consumption over the last 15 years. *Source* WEC (2016a)

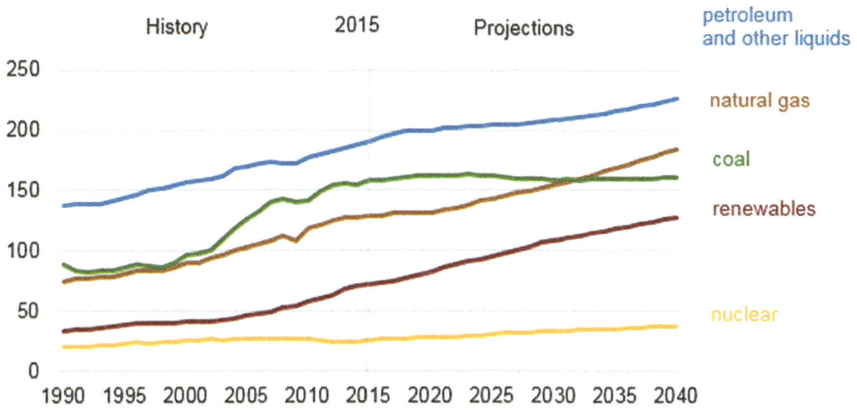

Fig. 1.2 Projected world energy consumption by energy sources, from 1990 to 2040, in quadrillion Btu. *Source* U.S. Energy Information Administration (EIA) (2017)

Regarding the other sectors, transport consumes approximately 27% of the world energy demand, accounting for 14% of emissions; the industrial sector accounts for 28%, emitting about 21% of greenhouse gases; and the construction sector consumes approximately 34% of the energy demand, emitting only 6% of GHG. Thus, considering that the urbanization process occurs in an increasingly intense manner worldwide, the eminence of developing technologies and innovation in the use of renewable sources in urban areas can be perceived, which represent key-parts to achieving long-term sustainability in the global energy system.

1.1.1 Non-renewable Energy

Also known as "conventional" sources of energy, "non-renewable" sources are those that are limited, that take a long time to replenish, and are not geographically distributed in a homogeneous manner. As mentioned earlier, these sources have historically dominated the world energy scenario, especially in relation to fossil fuels. Though, it should be clarified that a "non-renewable" energy source is not necessarily an "unclean" energy source, in terms of CO_2 emissions, among other greenhouse gases. The exception is nuclear energy, which is not considered renewable, but its utilization emits considerably less GHG than fossil fuels, although there are caveats regarding radioactive waste generated as a result of its use.

Fossil fuels originate from ancient matter, which has been subjected, from millions to hundreds of millions of years, to high temperatures and pressures inside the earth (Dunlap 2015). The result of this formation process, depending on the details of the formation condition or starting material, can be a solid (coal), liquid (oil) or gaseous (natural gas) fossil fuel.

Coal

Coal is an organic rock, formed from terrestrial plant matter that has accumulated in the bottom of a swamp for thousands of years. Coal can be found in different varieties, depending on age and formation condition. The younger type of coal, with the lowest calorific potential, is known as lignite. If the lignite is subjected to higher pressures and temperatures, the sub-bituminous coal is formed. With additional heat and pressure, the most abundant type of coal, bituminous, is formed. Finally, there is the anthracite coal, which is rarer, but has a high calorific potential. This solid fuel can produce energy through combustion, and, because it is quite abundant, it is still one of the cheapest types of fuel.

Coal was widely used initially in industrial processes and in the domestic environment. However, since the use of petroleum and its derivatives became popular, the use of coal in industry has been drastically suppressed, with the exception of the metallurgical industry. The coal combustion produces ashes, carbon dioxide (CO_2), sulfur dioxides (SO_2) and oxides of nitrogen (NOx), in larger quantities than those produced in the combustion of other fossil fuels. In addition, the surface mining of coal is known to cause several environmental problems, such as deforestation, water and soil contamination and erosion (Hinrichs et al. 2010).

Natural Gas

Natural gas is a gaseous mixture of light hydrocarbons, which can be extracted from deposits associated with oil or from deposits where it occurs on its own (Dunlap 2015). Like oil, it was formed over millions of years ago, from the decomposition of organic ancient matter. In terms of pollution risk, natural gas is the cleanest fuel among the fossil and its market is expanding rapidly, being expected to continue to grow in the next two decades as a result of the gas rush for energy production.

Natural gas is considered, regarding the chemical aspect, the simplest of fossil fuels, because it is mostly composed of methane gas (CH_4), containing fewer impurities than heavier hydrocarbons, making it burn more efficiently and producing less pollution. It has several practical industrial and domestic uses, like on heating and transportation, as boiler fuel and as raw material for the chemical industry (ammonia production, tensioactives, fertilizers, plastic, etc.) (Hinrichs et al. 2010).

However, although natural gas is relatively inexpensive to extract and is considered "cleaner" than oil and coal, when burned, it also releases significant concentrations of carbon dioxide. Moreover, extracting natural gas is also responsible for environmental problems. For instance, the process of fracturing rocks can cause mini-earthquakes and the high-pressure water and chemicals that are forced underground may leak to other sources of water, leading to contamination of ground and drinking water resources.

Petroleum

Petroleum, or Oil, is a mixture of crude oil, natural gas in solution and thick and heavy asphaltic semisolids. Petroleum is a complex mixture of hydrocarbons, having a mass average ratio of 1 hydrogen to 7 carbons. This mixture was formed through the decomposition of organic matter from marine fauna, over millions of years, under high pressures and temperatures related to deep burial.

Oil must go through a series of stages during the refining process to be converted into a useful product. Figure 1.4 shows a schematic representation of an oil fractionating distillation column. Considering that oil is composed by different hydrocarbons with different structures, each of these components has their own physical and chemical characteristics, being used in different applications. The ones with lager number of carbon atoms per molecule have higher boiling points, and this property is the basis of the refining process, as represented in Fig. 1.3. Gasoline is considered the most important petroleum derivative, accounting for about 45% of the production of a typical American refinery.

Although oil provides us with thousands of conveniences, the intensive use of this resource presents several negative environmental impacts, from its extraction to its final use. During oil extraction, the process of drilling may lead to oil spills, if

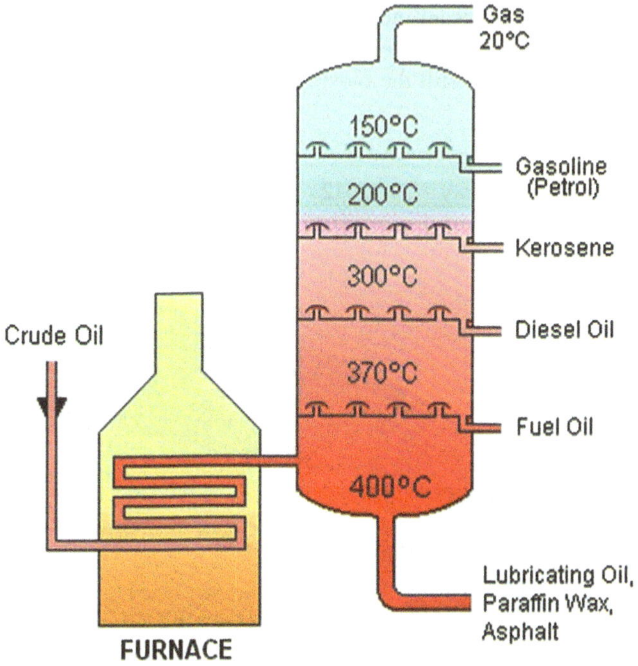

Fig. 1.3 Schematic representation of an oil fractionating distillation column. Based on Dunlap (2015)

there is a problem with the drilling machinery, causing explosions out of the well and spill into the surrounding land or ocean. Large oil spills are rare but can result in significant and long-term environmental adverse impacts, because spilled oil may be distributed into a number of different habitats and in different forms (IPIECA 2015). Moreover, during its use, the combustion of oil results in the emission of different pollutants, especially greenhouse gases, just like the other fossil fuels.

1.1.2 Climate Change

Over the last 150 years, fossil fuels have been and remain major components of our energy production. During this period, a substantial fraction of these reserves was exploited. As they are neither renewable nor clean sources, our significant dependence on these fuels, their depletion, and their considerable impact on the environment, lead to the crucial need for implementing renewable energy sources and diversifying the world energy matrix.

The consumption of these sources is particularly responsible for the anthropogenic emission of GHG in the last centuries, and indicated as accountable for climate change and global warming (WEC 2016a). Figure 1.4 shows the historical emission of GHG due to the use of fossil fuels since the 1990s, and how future projections foresee a significant increase over the next decades.

As they are significantly cost-effective and easy to store, pipe and transport everywhere, fossil fuels are still the most valuable source of energy. However, the

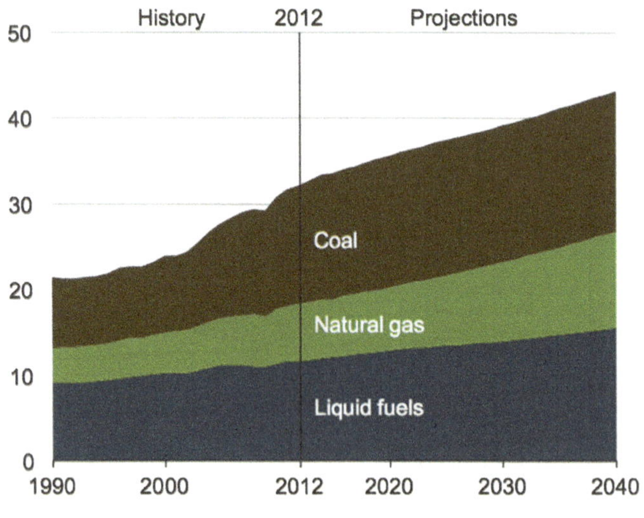

Fig. 1.4 Global CO_2 emissions related to the consumption of fossil fuels by fuel type, in the period of 1990–2014, in billions of tons. *Source* International Energy Agency (2016a, b)

exploitation of these resources has been proved harmful for the environment and they are very limited to fulfill our energy demands in future horizons.

Considering that the carbon in fossil fuels has been stored underground for millions of years, by removing this sequestered carbon from the bottom of the Earth and releasing it into the atmosphere, turns Earth's carbon budget out of balance. Carbon dioxide is a gas responsible for keeping heat in Earth's atmosphere, resulting in the "greenhouse effect" process. This effect is essential to life on Earth, but, when the carbon budget is unbalanced, it contributes to temperatures rising faster than organisms can adapt, aggravating climate change and global warming.

According to Dunlap (2015), over the past century, climate change has been manifesting in several ways, like in the reducing size and number of glaciers; the reducing area and thickness of the Artic sea ice; the increasing sea level; the biological changes in bird migration and stages of plant development; the increasing geographical range of certain plants and animals; the thawing of the permafrost in the Artic; and weather changes, like more frequent El Niño events. The release of GHG into the atmosphere by anthropogenic activities is the main cause of these changes, according to the general scientific consensus. Moreover, other anthropogenic factors, like deforestation, have also been contributing to the aggravation of the problem.

1.1.3 Renewable Energy

Also known as alternative sources of energy, "renewable sources" are those that do not rely on resources that are admittedly limited and whose utilization does not cause their depletion and exhaustion. Since alternative energies come from cycles for the conversion of solar radiation, the primary natural source of much of the available energy on Earth (Pachcco 2006), they are generally viewed as less impacting than fossil fuels. These sources can be explored in their most diverse forms of energy sources, such as solar, wind, hydropower, geothermal, tidal, ocean and biomass.

According to IPCC (2012), these alternative sources present not only the potential to slow down the effects of global warming, but can also offer much broader benefits. If these sources are properly implemented, they can contribute to economic and social development, to greater access to energy, to a safer electrical system, and to reduce the negative impacts of the energy system on the environment and the health of the population. Their use should also impact social development by creating an emerging industrial sector and, consequently, jobs and wealth.

As a result of the global quest for a more sustainable and less harmful global energy grid, the development of and investment on renewable energy systems increase significantly each year. As of 2015, renewable energy provided around 19.3% of global final energy consumption, experiencing continuous growth in capacity and production in 2016 (REN21 2017). Although the growth of renewables in the heating and cooling and transport sectors was relatively slow in 2016, the power sector was the one with the greatest increases in renewable energy capacity. According to REN21 (2017), the developing countries are the ones responsible for the installation of most of the new alternative energy capacity, especially China, starring as the single largest developer of renewable power and heat over the past eight years.

Table 1.1 represents the ranking of the Top 5 countries regarding total renewable energy capacity or generation as of end-2016. China, the United States, and Brazil are the three top countries in total installed renewable energy power, when hydropower is taken into consideration. However, as Brazilian energy grid is highly dependable of hydropower, and the sustainability of this source is greatly discussed, as it will be further seen in Chap. 2, when the total power does not include hydropower, the Top 3 countries regarding installed renewable energy power are China, the United States and Germany. China was among the Top 3 countries in almost all renewable energy sources considered, which is a result of the great attention and investment the country has being directing towards energy sustainability, in order to be able to fulfill their great energy demand.

Although China continues to be the leading country in alternative energy development, in 2016, renewable energy also spread to a growing number of other developing and emerging economies, some of which have become important

Table 1.1 Top 5 Countries: total capacity or generation as of end-2016

Ranking	1	2	3	4	5
Renewable power (incl. hydro)	China	United States	Brazil	Germany	Canada
Renewable power (not incl. hydro)	China	United States	Germany	Japan	India
Renewable power capacity per capita (not incl. hydro)	Iceland	Denmark	Sweden/Germany		Spain/ Finland
Bio-power generation	United States	China	Germany	Brazil	Japan
Geothermal power capacity	United States	Philippines	Indonesia	New Zealand	Mexico
Hydropower capacity	China	Brazil	United States	Canada	Russian Federat.
Hydropower generation	China	Brazil	Canada	United States	Russian Federat.
Concentrated solar power capacity	Spain	United States	India	South Africa	Morocco
Solar PV capacity	China	Japan	Germany	United States	Italy
Solar PV capacity per capita	Germany	Japan	Italy	Belgium	Australia/ Greece
Wind power capacity	China	United States	Germany	India	Spain
Wind power capacity per capita	Denmark	Sweden	Germany	Ireland	Portugal

Source REN21 (2017)

Table 1.2 Emission of greenhouse gases (CO_2) in electricity generation by different energy sources

Energy resource	CO_2 (kg/MWh)
Coal	955
Natural gas	430
Solar photovoltaic	98–167
Wind	7–9
Geothermal	7–9
Hydroelectric	3.6–11.6

Source Adapted from Dunlap (2015)

markets. Considering that more than 1 billion people do not have access to electricity, distributed renewable energy projects, especially those in rural areas far from the connected grid, can be an important, and often cost effective, option to provide such access. Moreover, the renewable energy sector is also a great source of employment. This sector employed over 9.8 million people in 2016, which represented an increase of 1.1% over 2015. Regarding the different technology, as reported by REN21 (2017), solar PV (photovoltaic) and biofuels provided the largest numbers of jobs. By region, employment shifted further towards Asia, which accounted for 62% of all renewable energy jobs, not including large-scale hydropower (REN21 2017).

The concept of clean energy is generally associated with renewable sources because, compared to fossil fuels, they have low environmental impacts and practically do not produce waste or the emission of pollutants (Villalva 2015). However, while all renewable energies produce less CO_2 per unit of electrical energy generated, none of them are actually zero carbon dioxide emissions.

Table 1.2 shows GHG emissions (kg CO_2) related to the production of 1 MWh (megawatt-hour) of energy through renewable and non-renewable sources. Through this analysis, it can be seen that, although far less significant than the emissions attributed the use of fossil fuels, renewable sources cannot be considered "carbon free".

In addition to the emissions, the implementation of systems for using renewable sources can also cause different impacts, such as the flooding of large areas in the case of hydropower, or generating audible noise and modifying landscapes in the case of wind energy. However, in general, despite the impacts, renewable energy sources are considerably much cleaner and safer compared to the conventional sources.

1.1.4 Challenges for the Development of Renewable Energies

Although we are on the path to transitioning from the global energy matrix to a more sustainable and less impacting system, the effective use of renewable energy sources still faces challenges. According to the IPCC (2012), notwithstanding the remarkable progress in the sector, the pace of improvement towards the use of cleaner energies is much slower than that needed to meet targets for significant emission reductions. Dunlap (2015) points out the following main challenges for the sustainable development of energy use: **availability of the necessary resources; economic and technological factors; socio-environmental factors; political factors;** and **integration of systems**.

The availability of the necessary resources relates not only to the energy source itself, but also to other resources needed to exploit it. The possibility of exploring a particular renewable source greatly depends on the location and amount of available energy, as well as the availability of systems, devices or machines capable to exploit it. Although sources of energy such as solar and wind are almost in every part of the world, there are also sources that are dependent on very specific conditions and can only be explored punctually, such as tidal and geothermal energy. In addition, in the case of using solar energy, for example, it requires producing solar panels, which require a large quantity of the following chemical elements: silicon, copper, indium, gallium and selenium. These elements are relatively rare and expensive, in addition to the fact that their currently produced quantity is thousands of times less than what would be needed to produce enough solar cells to meet the world's energy consumption needs (Bleiwas 2010).

In addition to the limitations related to the availability of resources, the use of renewable sources also encounters some technological and economic barriers. The present technological development of renewable energies systems aims at achieving higher quality, lower cost, and efficiency in the production and also in the application of such energy resources. However, although there is appropriate technology to exploit a certain energy source, this alone is not sufficient to ensure that its utilization is environmentally or economically viable. For instance, in the case of Brazil, although there is a large potential for exploring renewable energy sources, there are still market barriers to be overcome for the commercial implementation of several renewable energy technologies (Worldwide Fund for Nature (WWF) 2012). The production scale of these technologies poses major challenges to the effectiveness of its implementation, due to the small size of industrial parks and a restricted market, causing the installation costs to remain high.

The development of renewable energy projects is also responsible for negative social and environmental impacts, as mentioned previously, a factor that in some cases prevents or delays its implementation. In addition, public acceptance still poses a challenge, regardless of the energy source involved, mainly due to the "not in my backyard" attitude towards implementing energy initiatives (IPCC 2012). The main socio-environmental impacts related to the energy sources that are presented in this work will be commented in more detail later.

As regards political factors, the implementation of many renewable initiatives still depends on political and economic incentives. Public investment policies in the renewable energy sector can help combat market barriers and lower electricity production costs. In addition, tax and credit incentives are essential for expanding the market and for more private investment in the sector. Goals to increase the participation of these sources in the energy matrix should also be established in regional, state and federal planning so they are effectively implemented.

Finally, the advantages in using renewable sources are widely recognized. However, its inconsistent and often unpredictable nature is seen as barriers for large-scale integration into the conventional energy grid. Some of these challenges are: fluctuations of the energy generated; reduced assurance of energy and power; impact on voltage regulation in the local network; and the stability of the electrical system. The particularities of energy systems represent a large part of the problem, since they need to balance supply and demand relationships due to the still high cost of electricity storage systems.

1.2 Brazilian Energy Matrix

The Brazilian energy matrix has a unique characteristic that differentiates it from the rest of the world: it has a significant share of renewable energy sources. Particularly with respect to its electricity generation, Brazilian matrix is characterized by predominance in the share of hydroelectric energy to generate electricity.

Figure 1.5 shows data which indicate that their national energy matrix is currently composed of about 41% of renewable sources, such as sugarcane biomass (16.9%), hydroelectric power (11.3%), and firewood and coal (8.2%), among others (4.7%). Non-renewable sources account for about 59% of the matrix, including petroleum (37.3%), natural gas (13.7%), coal (5.9%), uranium (1.3%), among others (0.6%) (MME Brasil 2016).

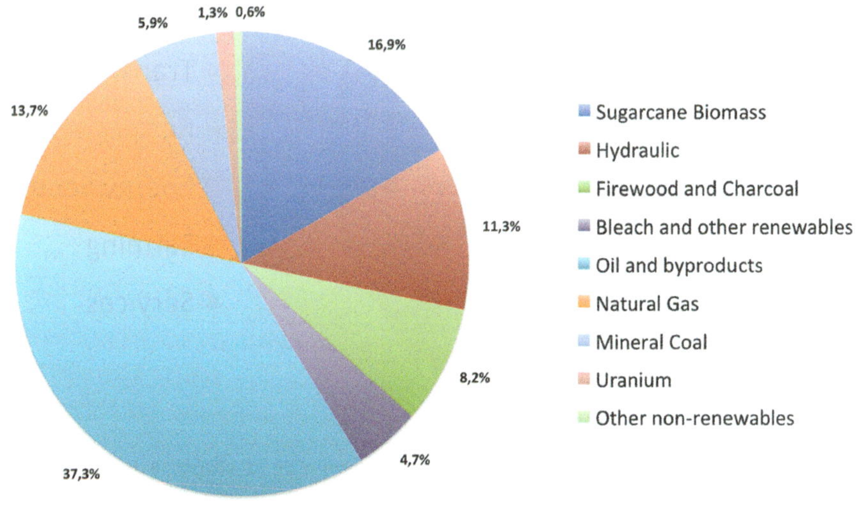

Fig. 1.5 Brazilian energy matrix in 2015. *Source* Adapted from MME Brazil (2016)

Among the "black liquor and other renewables" category, responsible for 4.7% of the national energy, the following partition of the different sources that compose it can be observed: 55.7% black liquor, 22.0% biodiesel, 13.1% wind, and 8.1% of other biomasses, in addition to solar energy, biogas and industrial vegetable coal gas, contributing less than 1% in the category (MME Brasil 2016). Notwithstanding the still modest participation, solar, wind and biogas sources have been gaining ground in the Brazilian market in recent years, demonstrating they are increasingly important for the national energy issue.

In Brazil, in 2015, the domestic energy supply reached a total magnitude of 300 megatons of oil equivalent (Mtoe) and around 261 Mtoe were consumed, in which industrial production, cargo transportation and mobility accounted for approximately 65% of that consumption (MME Brasil 2016). Figure 1.6 shows the distribution of each sector`s share in the consumption of domestic energy offered in Brazil in 2015.

In the industrial sector, electricity, sugarcane bagasse, mineral coal and natural gas are underscored, with renewable energy consumption accounting for 54% of the total in 2015. In the transport sector, fossil fuels, such as gasoline and diesel, are the most widely used, although ethanol is also underscored. As for residential consumption, despite an increase in the use of firewood and liquefied petroleum gas (LPG), electricity still leads in this consumption, with 45.2% of the total in 2015, but with a decrease of 0.7% when compared to 2014 (MME Brasil 2016).

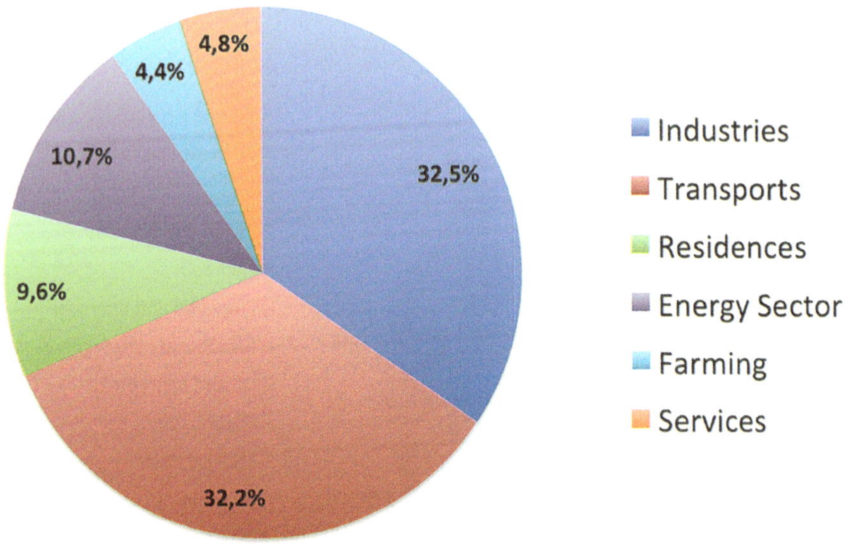

Fig. 1.6 Distribution of each sector's share in the consumption of energy in Brazil in 2015. *Source* Adapted from MME Brazil (2016)

1.2.1 *Retrospective of Production and Consumption*

The use and production of electricity in Brazil began in the late nineteenth century, when in 1883 the first hydroelectric plant was constructed in the city of Diamantina. That same year, D. Pedro II, granted Thomas Edison permission to introduce electric lighting equipment in the country, inaugurating the first municipal public electric utility service in Brazil in the city of Campos (Jannuzzi 2007).

Throughout the 20th century, Brazil witnessed intense economic development, which triggered an increasing demand for primary energy and the system growth. According to Tolmasquim et al. (2007), a considerable industrialization process is among the main factors that determined this growth, such as the installation of energy intensive plants, and a significant demographic expansion, accompanied by a rapid increase in the rate of urbanization.

At the beginning of the 21st century, Brazil experienced the greatest electricity crisis in its history, accentuated by water crises in some regions, which resulted in rationing programs in various parts of the territory and the institution of the Chamber for Management of the Electricity Crisis (GCE). From 2000 to 2010, several power plants were built for hydroelectric, thermoelectric and biodiesel. In addition, new companies and programs such as the 'Light for All' and PROINFA (Alternative Energy Sources Incentive Program) were created to increase the share of electricity production based on wind, biomass and small power plants in the National Interconnected Electrical System (SIN).

The model to nationalize energy projects, established in the 1940s and 1960s, meant that much of the production and distribution of electricity, oil, and gas was under the responsibility of state and federal governments, previously dominated by foreign-owned projects (Gomes and Vieira 2009). However, although the model was considered successful until the 1980s, it gave rise to problems in the Brazilian energy matrix, such as the artificially low tariffs for electricity and the construction of hydroelectric plants for political benefits, with no guaranteed return with good monetary rates (Goldemberg and Lucon 2007).

1.2.2 *Electrical System Characteristics*

The Brazilian electricity system can be divided into three different components: Generation, Transmission and Distribution. Regarding generation, currently, more than 75% of the electricity matrix originates from renewable sources, with hydroelectric projects accounting for almost 80% of the entire generation in 2016 (ANEEL 2016b), as shown in Fig. 1.7.

Although the source with the greatest impact on the expansion of Brazil's electric power matrix continues to be the water source, in the last decade the country has presented considerable growth in the share of other renewable sources, mainly in relation to wind energy and biomass, which represent, respectively, 5%

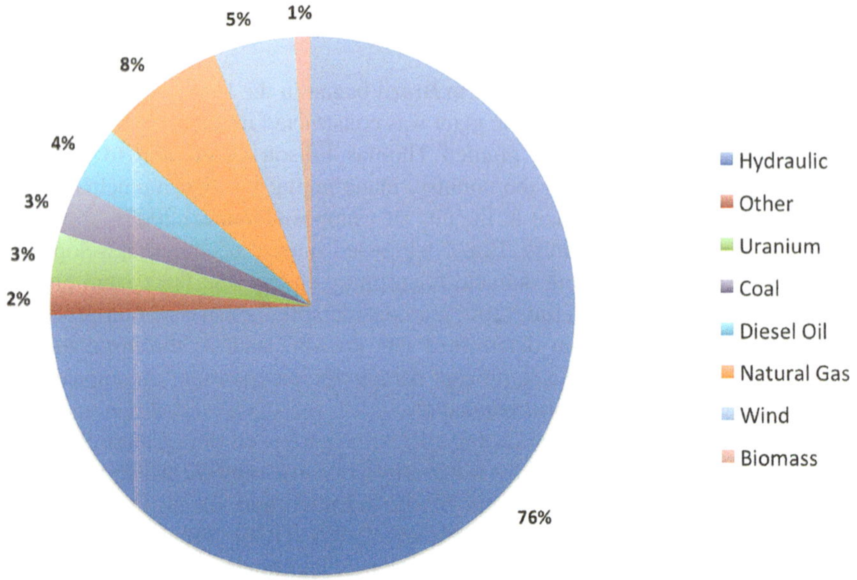

Fig. 1.7 Share of each electricity generating source in total—Jan–Sep/2016. *Source* ANEEL (2016b)

and 1% in the generation of the country's electricity (ANEEL 2016b). Up to October 2016, more than 2000 MW (megawatt) of installed capacity of Wind Power Plants (WPP) entered commercial operation and about 350 projects are under implementation, with an estimated 350 MW for operation start-up and more than 8,000 MW between 2017 and 2021 (ANEEL 2016b).

Regarding transmission, according to ANEEL (2008), there are more than 90 thousand kilometers of transmission lines across Brazil. These are operated by 64 concessionaires, which are responsible for the implementation and operation of the network, which has a large extension particularly due to the distant location of the generation sources to large consumer centers. The Brazilian network is divided into two large blocks: The National Interconnected System, which covers a large part of the territory, and the Isolated Systems, predominant in the northern region (ANEEL 2008).

The National Interconnected System (SIN) is operated and coordinated by the National Electric System Operator (ONS) and is composed of companies from the South, Southeast, Midwest, Northeastern and part of the North, presenting 96.6% of the country's electricity production capacity (ANEEL 2008), which is a system with unique size and characteristics, with a strong predominance of hydroelectric plants. The Isolated Systems serve the states of Amazonas, Roraima, Acre, Amapá, Rondônia and Mato Grosso, as well as the island of Fernando de Noronha, which is predominantly supplied by fossil fuel-fired thermal power plants, as well as by small hydroelectric power stations, hydroelectric and thermoelectric power stations fueled by biomass. The existence of these isolated systems is due to the region's remote location, which leads to some logistics supply difficulties (ANEEL 2008).

Finally, the distribution of electricity is carried out by distributors, which are the main ones in charge of serving the consumer public. According to ANEEL (2008), the country has about 63 concessionaires, state or private, servicing more than 61 million consumer units. The distributors are responsible for receiving the load from the transmission supply companies of the country, creating a bridge between the plants and the society. In addition, they are also responsible for transforming the received load into variable voltages at a lowered and stabilized voltage for residential use.

The activities developed by the distributors are regularized and controlled by ANEEL, through indicators such as DEC (Equivalent Interruption Duration per Consumer Unit) and FEC (Equivalent Interruption Frequency per Consumer Unit). The quality of the service provided and the tariffs charged are some variables that are also regulated, and the rates are charged according to the amount of electricity consumed by the residence in the previous month, with the addition of charges and taxes by the electric sector determined by law (ANEEL 2008).

1.2.3 Programs and Incentives

The implementation of public investment policies in the renewable energy sector in Brazil can reinforce the fight against market barriers and reduce the costs of producing electricity. Tax and credit incentives are essential for expanding the market and private investments in the sector (WWF 2012). Among the national legal incentive mechanisms, Proinfa, instituted by Law 10,438 of April 26, 2002, is considered one of the most important Brazilian experiences for encouraging renewable energies.

The program began two years after its creation, aimed at increasing the participation of alternative sources in the National Interconnected System. Its planning foresees a two-stage undertaking, although the second stage is currently delayed and with no specific starting date. In the first stage, the program encouraged the development of renewable energy sources (wind, biomass and small hydroelectric plants or SHPs), in order to increase its market competitiveness, to generate large-scale gains, promote technological learning and assimilate the environmental, social, technical and economic benefits of projects to generate renewable sources. Eletrobrás is responsible for purchasing the energy produced with the support of the program following the feed-in tariff mechanism. In addition to this mechanism, Proinfa introduces other mechanisms to subsidize investments, providing special credit lines from the National Economic and Social Development Bank (BNDES) for the selected projects.

Proinfa stipulates, as a goal of the first phase, the implementation of an installed capacity of 3.29 million kW (kilowatt) produced by 144 renewable energy plants—63 SHPs, 54 wind farms and 27 biomass plants. In addition, 60% of the construction costs of the projects must be nationalized, contributing to the insertion of small energy producers in the sector and their diversification. BNDES has created credit lines to cover up to 70% of the investments. Although not available on the total invested by BNDES in Proinfa, in the first quarter of 2011, US $ 650 million was invested by the bank energy sector, of which US $ 290 million was invested in alternative sources (WWF 2012).

References

Agência Nacional de Energia Elétrica (ANEEL) (2008) Atlas de energia elétrica do Brasil, 3rd edn. Brasília

Agência Nacional de Energia Elétrica (ANEEL) (2016b) Boletim de Informações Gerenciais 2016. Available at http://www.aneel.gov.br. Accessed 27 Feb 2017

Araújo MZT, Vieira VPPB (2001) As dificuldades encontradas para implantação da Barragem do Castanhão e o acompanhamento das ações mitigadoras dos seus impactos ambientais. XXIV Seminário Nacional de Grandes Barragens, Fortaleza

Bleiwas DI (2010) Byproduct mineral commodities used for the production of photovoltaic cells. Geological Survey Circular 1365. Available at http://pubs.usgs.gov/circ/1365/. Accessed 4 Aug 2017

Dunlap RA (2015) Sustainable energy. Dalhousie University, Halifax

Goldemberg J, Lucon O (2007) Energia e meio ambiente no Brasil. Estudos Avançados 21(59):7–20

Gomes JPP, Vieira MMF (2009) O Campo da energia elétrica no Brasil de 1880 a 2002. RAP: revista de administração pública, Rio de Janeiro 43(2): 295–321. Available at http://www.scielo.br/pdf/rap/v43n2/v43n2a02.pdf. Accessed 3 Aug 2017

Hinrichs RA, Bach M, Dos Reis LB (2010) Energia e meio ambiente. Cengage Learning, São Paulo

Intergovernmental Panel on Climate Change (IPCC) (2012) Renewable energy sources and climate change mitigation: special Report of the Intergovernmental Panel on Climate Change. Cambridge University Press, New York

IPIECA (2015) Impacts of oil spills on marine ecology: good practice guidelines for incident management and emergency response personnel. London

Jannuzzi AC (2007) Regulação da Qualidade de Energia Elétrica sob o Foco do Consumidor. Dissertação de Mestrado em Sistemas Elétricos de Potência, Publicação PPGENE.DM - 302A/07, Departamento de Engenharia Elétrica, Universidade de Brasília, Brasília

MME Brasil (2016) Balanço Energético Nacional 2016. EPE, Brasília. Available at http://www.cbdb.org.br/informe/img/63socios7.pdf. Accessed 7 Aug 2017

Pacheco F (2006) Energias renováveis: breves conceitos. Conjuntura e Planejamento Salvador 149:4–11. Available at http://ieham.org/html/docs/Conceitos_Energias_renováveis.pdf. Accessed 27 Feb 2017

REN21 (2017) Renewables 2017 global status report. REN21 Secretariat, Paris

Tolmasquim MT, Guerreiro A, Gorini R (2007) Matriz Energética Brasileira: uma prospectiva. CEBRAP, São Paulo. Available at http://www.scielo.br/scielo.php?pid=S0101-33002007000300003&script=sci_arttext. Accessed Aug 4 2017

U.S. Energy Information Administration (EIA). International energy outlook 2017—with projections to 2050. Available at www.eia.gov. Accessed 25 Dec 2017
Villalva MG (2015) Energia solar fotovoltaica: conceitos e aplicações, 3 edn. Érica, São Paulo
World Energy Council (WEC) (2016a) World energy resources 2016. London
Worldwide Fund for Nature (2012) Sumário para tomadores de decisão: Além de Grandes Hidrelétricas. Políticas para fontes renováveis de energia elétrica no Brasil. Brasilia. Available at http://d3nehc6yl9qzo4.cloudfront.net/downloads/alem_de_grandes_hidreletricas_sumario_para_tomadores_de_decisao.pdf. Accessed 24 Mar 2017

Chapter 2
Renewable Energy Sources

Abstract The content describes the most growing renewable energy sources, i.e., hydropower, wind and photovoltaic. It shows the basic principles for each generation technology, its sustainability aspects, the world availability and presents their potential and generation in Brazil.

As already defined in Chap. 1, renewable energy sources are those which, over relatively long horizons of time, come from resources that are not recognized as being limited and whose exploitation does not cause depletion or exhaustion. This chapter seeks to provide more details on the three energy sources considered sustainable that stand out most in the international context: hydropower, wind energy and photovoltaic solar energy. Therefore, the chapter presents the historical context of development, the international scenario, the Brazilian energy potentials, and the characteristics of exploitation for each of these sources, in addition to a discussion related to the dimensions of their sustainability.

2.1 Hydro-electric Energy

The mechanical power of falling water is an age-old resource used for carrying out services and for productive use. More than 2000 years ago, the Greeks used vertical and horizontal axis machines and water wheels to grind wheat into flour (Hinrichs et al. 2010). In the sixteenth century, this energy was widely used for milling and pumping. The 19th century brought, in addition to the development of the steam engine, the major innovation of electricity generators. This new technology enabled to install the first hydroelectric plant at Cragside, Rothbury, England, in 1870. However, industrial use of hydroelectric power began only a decade later in 1880 in Grand Rapids, Michigan, when a dynamo driven by a water turbine was used to provide lighting for a theater and storefront (IPCC 2012).

According to the World Energy Council (WEC) (2016a, b, c), hydropower is the main renewable power source, providing about 71% of all global renewable energy

J. F. G. de Oliveira and T. C. G. Trindade, *Sustainability Performance Evaluation of Renewable Energy Sources: The Case of Brazil*,
https://doi.org/10.1007/978-3-319-77607-1_2

by the end of 2015. The undeveloped global potential is of approximately 10,000 TWh/year, with global hydropower capacity increasing by more than 30% between 2007 and 2015, totaling 1209 GW (gigawatt) in 2015 (WEC 2016a). As a renewable source, hydropower offers significant potential for GHG emission reductions. In addition, the technology for hydroelectric development is technically mature, already economically competitive with current energy prices in most markets, which means that it is still expanding at an accelerated pace. This significant increase in hydropower capacity over the last 10 years is anticipated, in many scenarios, to continue in the short term (2020) and medium term (2030), raising a number of environmental and social concerns which are perhaps the major challenges for the continuous development of this technology (IPCC 2012).

2.1.1 World Hydroelectric Generation

Since the 20th century hydropower has been essentially used as a source of electricity. This resource has been and continues to be the most popular method of generating electrical energy that is generally considered renewable and carbon free. The extent to which this qualification of hydroelectric power is accurate depends mainly on how this resource is utilized and includes factors such as geography and climate, as will be discussed in the Sect. 2.1.3. Although big hydropower projects were known to cause significant environmental impacts during their construction, there has been progress in terms of sustainability practices in the sector, causing an increase in the acceptance by external stakeholders, such as non-governmental organizations (NGOs) and the financial community, which had previously opposed the development of some new projects (WEC 2016c).

Considered a very robust and consolidated technology, in recent years, hydropower has been experiencing globally a major upsurge in development, providing today over 16% of global electricity production. The total installed capacity of this resource has grown by 39% from 2005 to 2015, with an average growth rate around 4% per year. This rise has been specially concentrated in emerging markets. For them, hydropower offers not only clean energy, but also water services and energy security, facilitating regional cooperation and economic development. The main drivers for the upsurge in hydropower development include: the increased demand for electricity, energy storage, flexibility of generation, freshwater management, and climate change mitigation and adaptation solutions (WEC 2016c).

At the end of 2016, the leading hydropower generating countries were: China, Brazil, Canada, United States, India and Russia. As represented in Fig. 2.1, China leads the ranking, producing almost 1127 TWh (96.9 Mtoe) of hydroelectricity per year. With roughly a third of this annual production, Brazil comes in second, generating annually more than 382 TWh (32.9 Mtoe) of hydroelectricity, followed closely by Canada, in third place, with a production of almost 376 TWh (32.3 Mtoe) per year.

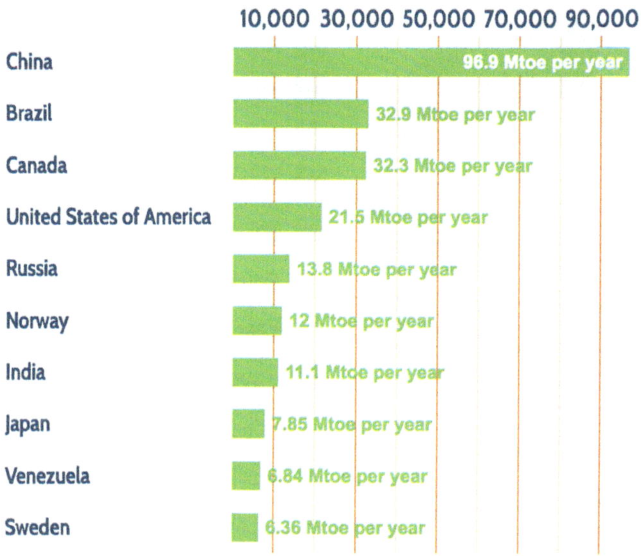

Fig. 2.1 Top hydropower producing countries in 2016. *Source* WEC (2017)

Considering hydroelectricity production by region, East Asia comes in first place, with 381 GW of installed hydropower capacity, followed by Europe (293 GW) and North America (193 GW), as represented in Table 2.1. The Top 3 regions regarding hydropower capacity represent more than 72% off all hydropower capacity, as shown in Fig. 2.2. However, Latin America is considered a key market for hydropower development, due to its huge unexplored hydropower potential, and Africa is expected to be a major market for future hydropower activity (WEC 2016c).

In 2015, China has strengthened its leading position by adding more hydropower capacity, with new developments taking place at Baihetan (16 GW) and Wudongde

Table 2.1 Hydropower total installed capacity by region

Region	Hydropower installed capacity (GW)
East Asia	381
Europe	293
North America	193
Latin America and The Caribbean	159
South and Central Asia	72.3
South East Asia and Pacific	57.8
Africa	25.3
Middle East and North Africa	20.6

Source WEC (2017)

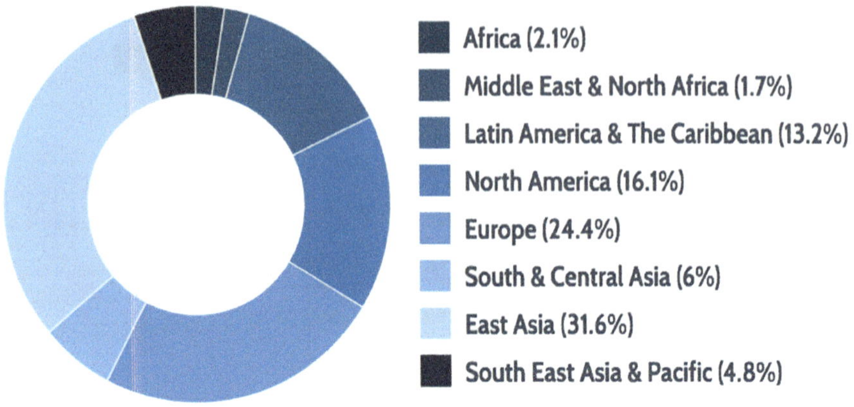

Fig. 2.2 Hydropower installed capacity by region. *Source* WEC (2017)

(10.2 GW). As reported by WEC (2016c), the Chinese total capacity is expected to reach 350 GW of pure hydropower plus 70 GW of pumped storage by 2020. However, China is not the only country in Asia that shows significant advances in hydropower development. New deployments have been taking place in the emerging Asian markets, including concentrations in Russia, India, Turkey and Vietnam (WEC 2016c).

Although Asia is considered as the next likely leading market for future development, Latin America is another key market for hydropower investments. Leading the continent in installed capacity and new capacity additions, Brazil already presents a total hydropower installed capacity of 91.8 GW. As it will be further addressed in the next sections, hydropower forms the backbone of Brazil's electricity system, supplying more than 60% of its electrical demand. Aside from Brazil, other Latin American countries with significant hydropower capacity include Argentina, Chile, Colombia, Paraguay, Peru, Venezuela and Ecuador (WEC 2016c).

As mentioned, another region expected to be a major market for future hydropower activity is Africa. In 2012, the continent showed the low average electrification rate of only 45%. Therefore, developing hydropower can represent a real opportunity for providing electricity on Africa using largely local or regional resources. According to WEC (2016c), to date, only 9% of the African reported hydropower potential is currently developed, showing that significant undeveloped remains across all of the continent, specially in the markets of the Democratic Republic of the Congo, Angola, Ethiopia and Cameroon. Moreover, regional African co-operation bodies, like the Eastern Africa Power Pool, the West African Power Pool and the Southern African Power Pool, show the potential to encourage further advancement of hydropower in regions where domestic resources can be developed for exportation to neighbouring countries with greater electrical demands (WEC 2016c).

2.1.2 Hydroelectric Potential and Generation in Brazil

In 2015, according to the Brazilian National Energy Balance (BEN), prepared by the Brazilian Energy Research Company (EPE), hydroelectric source energy accounted for 11.3% of the Brazilian energy matrix, being surpassed by sugarcane byproducts, oil byproducts and natural gas. With regard to the domestic supply of electricity, which totaled 615.9 TWh in 2015 (a reduction of 1.3% in relation to the previous year), hydroelectric power accounted for 64% of Brazilian electricity production, still the main source of electricity in the country (BRASIL 2016). By 2016, hydroelectric power contributed with 76% of the electricity generated (ANEEL 2016b).

This significant participation is not by chance, since Brazil is one of the countries with the highest hydroelectric potential in the world, with a total estimated capacity of 260 GW (ANEEL 2008). Of this total, just over 30% were, in fact, transformed into plants built or granted. According to ANEEL (2008), the potential that can still be exploited in the future is of approximately 126 GW, of which more than 70% is in the Amazon and Tocantins/Araguaia basins in the northern region of Brazil.

Figure 2.3 shows that only 30% of the hydroelectric potential of the region is in operation or under construction, in contrast to almost 60% of the South and Southeast regions. This great untapped potential is related not only to the topography of the country, but also to how the national hydroelectric park was developed (ANEEL 2008). The first major Brazilian hydroelectric plant began operating in the 1950s in the Northeast (Paulo Afonso I, with a capacity of 180 MW), by the S. Francisco Hydroelectric Company. The large hydroelectric plants that followed were erected over the next 60 years and concentrated mainly in the South, Southeast and Northeast regions. In the North, the Tucuruí mills in Pará and Balbina in the Amazonas are some examples. However, it was only in the 1990s that the region began to be more extensively exploited, with the construction of the Serra da Mesa (GO) plant on the Tocantins River (ANEEL 2008).

The use of this great remaining potential should be carefully studied, considering that the Amazon is the largest Brazilian biome, concentrating countless species of fauna and flora, already threatened by the deforestation. Also, due to the flatter region topography, the construction of a large hydroelectric plant with an accumulation reservoir in the region would represent the flooding of a very large area, resulting not only in the devastating loss of biodiversity, but also the displacement of riverine populations that could endure social, material and cultural impacts.

Figure 2.3 shows that, with much of its hydroelectric potential already being explored, the South and Southeast regions have the highest installed hydroelectric powers of about 41.7 and 43.9 GW, respectively. Next, we have the Midwest region, with a hydroelectric capacity of 39.5 GW, which represents about 35% of its potential, and the Northeast region, with a hydroelectric capacity of 22.1 GW, which represents about 53% of its potential. Finally, we have the North region, with installed potential of 11.4 GW.

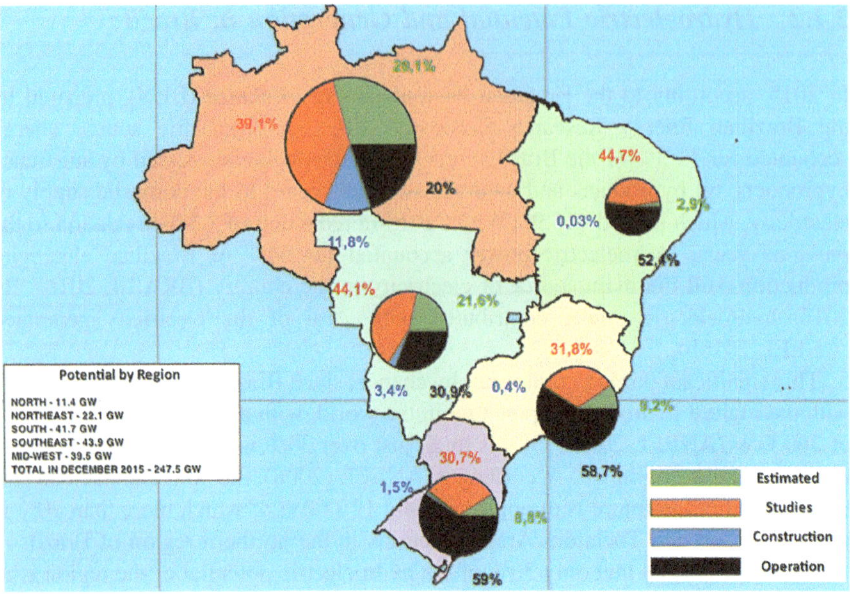

Fig. 2.3 Brazilian hydroelectric utilization by region. *Source* Eletrobras (2015)

2.1.3 Basic Principles of Hydropower Generation

The energy associated with running water can be harnessed in two ways: as high potential energy of the water in the dam is flown and used to rotate turbines at the base of the dam or as just the kinetic energy of running water with no reservoir. Figure 2.4 shows the first mode, explored by the construction of plants, generally a larger size, known as large falling water mills. The second mode is usually employed in plants known as "run-of-river water". In both manners, hydroelectric power is generated by using the water flow in a power plant, whose civil works, which involve both the construction and the diversion of the river and the formation of the reservoir, are as important, as the equipment installed (ANEEL 2008).

A hydroelectric plant is generally composed of a dam, a water intake, a powerhouse and a spillway, as can be seen in Fig. 2.4. In plants that take advantage of the potential energy of water, the construction of a dam is necessary to interrupt the normal course of the river and to allow the formation of the reservoir. This allows forming the necessary water level for the configuration of the hydropower, the adequate water abstraction and the regularization of the river flows during the periods of rain or drought (ANEEL 2008).

The capture and adduction systems consist of pipes that carry water to the powerhouse, where the turbines are, formed by a series of blades attached to an axis

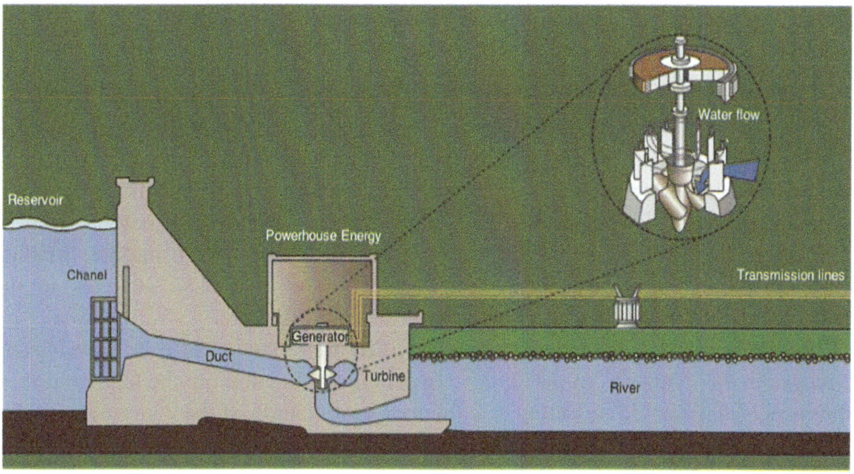

Fig. 2.4 Schematic profile of a hydroelectric plant with accumulation tank. *Source* ANEEL (2008)

connected to the generator. Water energy (kinetic) is used to move the turbines, whose spinning motion converts that energy into electricity by means of generators. Each turbine is adapted to operate in plants with a certain range of fall flow and height, and the main types of hydraulic turbines are: Pelton, Kaplan, Francis and Bulbo. Finally, there is the spillway, which allows the water outlet when the reservoir levels exceed the maximum limits.

In Brazil, the hydroelectric power installed in a power plant determines its classification. According to Law No. 13,097 of 2015, plants with installed capacity of up to 100 kW, and falling heights between 15 and 50 m, are classified as Micro-centrals. Plants with installed power from 100 kW to 3 MW, and falling heights between 20 and 100 m, are classified as Micro-generation plants. Plants with installed capacity of 3 MW at 50 MW, and falling heights between 25 and 130 m, are classified as Small Hydroelectric Plants (SHPs). Therefore, plants with installed power above 50 MW are classified as Hydroelectric Power Plants, or HPPs. Undertakings with an installed capacity of less than 3 MW (Micro generation plants) are exempt from any type of concession, consent or authorization, and must only be notified to the granting authority.

In Brazil, the greater the installed power of a plant, the farther away it tends to be from large consumer centers. Thus, transmission of the energy produced requires constructing high voltage power lines (from 230 to 750 kV), which can cross the territory of several states. The SHPs and smaller, installed in small falling water, generally supply small consumer centers and do not require such investment and operation of sophisticated facilities for energy transmission (ANEEL 2008).

2.1.4 Sustainability Issues of Hydroelectric Power

As already discussed, Brazil has great hydroelectric power production potential, reaching a prominent position among the largest hydropowers in the world. However, although it is produced through a renewable source that does not emit greenhouse gases, hydroelectric energy can cause significant impacts on the environment where it is generated. One of the main aspects responsible for these impacts is the construction of a reservoir for large plants, which implies flooding large areas and considerable physical-chemical and biological changes in the environment.

The main physicochemical impacts identified are reduction of river currents and changes in the dynamics of the aquatic environment. In addition, the temperature of the river is also altered, resulting in a thermal stratification of the dam, reducing the mixture of water from the dammed environment, creating anoxic conditions, favoring its eutrophication, and facilitating the occurrence of chemical reactions, which result in harmful compounds. With regard to biological impacts, there is a great loss of biodiversity and the establishment of a physical barrier for terrestrial and aquatic species constituted by the dam. The dam may also cause isolation of aquatic populations and prevent or hinder spawning (Sousa 2000).

The construction of a reservoir can also have negative effects on the local population and their traditional communities. This is due to the fact that the flooded area can lead to the loss of important elements of the natural, cultural and social patrimony attributed to it. The construction of a hydroelectric plant can also cause considerable population displacement, which causes this population to experience abrupt changes in how it interacts with the soil and in its relationship with its surroundings (Mendes 2005).

In addition to the impacts caused by the construction of the plant, there are also the others created during its operation, such as the formation of methane gas, due to the decomposition of the flooded organic matter, and also the gradual sedimentation of the reservoir. In Brazil, the vegetation is not adequately and satisfactorily removed during the cleaning of the area to be flooded. The decomposition of the vegetation that remains at the bottom of the reservoir generates methane gas, which has a high greenhouse effect potential, which is the origin of the emissions represented in Tables 1.1 and 3.1. In addition, the occupation of the banks of lakes and the increase of sediment deposition, as a result of the transformation of a previously lotic environment (running waters) into lentic (standing waters), lead to sedimentation of the reservoir and to the decrease of the plant's generation capacity.

A less impacting alternative to large hydroelectric plants for the utilization of hydroelectric power could be SHPs. These cause less impact to the environment, since they use the same renewable energy source, but operate on a smaller scale. In other words, they do not require the construction of extensive reservoirs, preventing such large impacts. Another advantage of SHPs is related to decentralized electricity generation in Brazil, because its greater technical and economic feasibility allows them to be installed in small centers previously not reached by the interconnected power grid.

2.2 Wind Energy

The use of wind power by man is also millenarian. Throughout its history of using wind power, in addition to being a driving force, wind has also played an important role in mechanical applications, such as grain grinding, water pumping and machinery (IPCC 2012).

There are indications of the presence of windmills in China and Babylon between 2000 and 1700 BC which were used to pump water and grind grains (Hinrichs et al. 2010). In the eleventh century, upon returning from their missions in the East, traders and crusaders were responsible for propagating the technology of these mills in Europe, which were used for food production in the Middle East. The Dutch then adapted this technology for industrial applications such as sawdust, papermaking and drainage of lakes and swamps (IPCC 2012).

However, electricity generation using wind power on a viable commercial scale began only in the 1970s thanks to technical advances and government support in countries such as Denmark, the United States, Germany and Spain (IPCC 2012). On the other hand, Brazil has been showing considerable growth in this sector over the last decades, with an impressive 114% increase in the production of electricity from wind source in the first half of 2015. In 2014, this matrix was responsible for only 1.4% of the total generated annually in the National Interconnected System (SIN). Currently, this share has grown, and wind energy already accounts for about 5% of all electricity produced in the Brazil (ANEEL 2016b).

2.2.1 World Wind Energy Generation

Just like solar energy, wind energy can be considered to be of virtually infinite longevity. However, to determine whether or not its utilization is feasible, it is imperative to consider the wind energy density available in each region. It is important to point out that the wind resource is not uniformly distributed throughout the planet, as represented in Fig. 2.5.

Moreover, a variety of other regional factors may restrict the growth of this type of exploitation, well before any limits of global technical and technological resources (IPCC 2012). As a result, the wind energy contribution to the national energy grid will differ for different countries. Considering that wind conditions in all locations are somewhat difficult to predict over a time horizon that exceeds a few days, this energy source presents considerable difficulty in providing a stable base of electricity for the electrical system. Therefore, in order for the electricity supply to be constant and to meet the needs of consumers, the production of wind energy

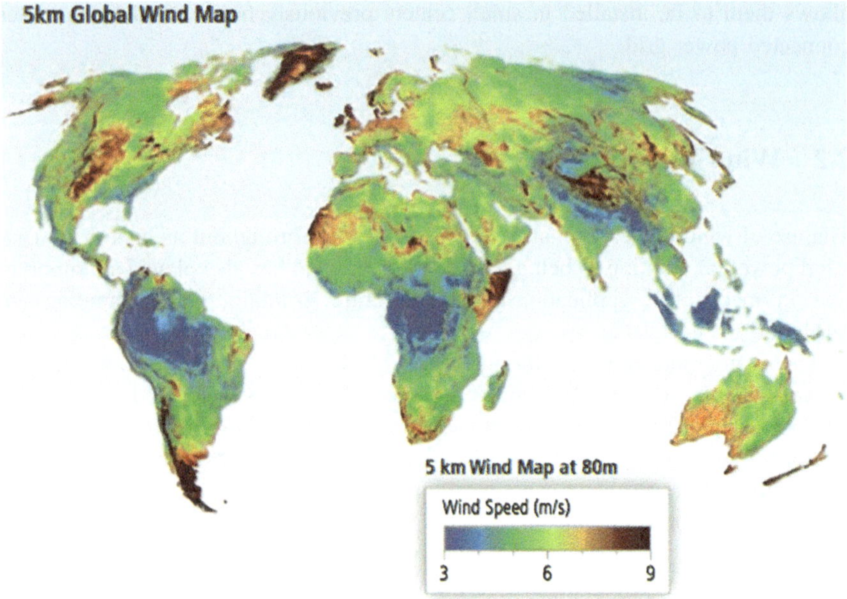

Fig. 2.5 Example of global wind resource distribution. *Source* IPCC (2012)

must be combined with that of other renewable sources, such as solar and hydropower.

World wind power generation capacity has reached 540 GW at the end of 2015, representing around 5% of total global power generation capacity, as shown in Fig. 2.6. Although the theoretical potential for wind energy, estimated by the annual global flow, is 6000 EJ/year, or about 1,7 million TWh/Year, (Rogner et al. 2000), the overall technical potential of wind power is not fixed, as it is related to the state of the available technology, among other restrictions related to the use of energy. Despite this, a growing number of assessments of global wind resources show that the world's technical potential for this type of energy exceeds current global electricity production (IPCC 2012).

Few countries have seriously invested in exploiting their wind potential, with Germany and Spain being notable in this regard. Denmark, on the other hand, stands out on a per capita basis of production, currently owning about 20% of wind power on its electricity matrix. This country has significantly developed offshore wind capacity. This technology is able to take advantage of the large wind power that offshore regions have and perhaps reduce some of the environmental concerns that technology presents (Dunlap 2015). Table 2.2 contains information on the current state of global wind energy utilization.

According to WEC (2016b), during 2015, the total investments in the global wind sector reached a record level of USD109.6 billion, and more than 93% of

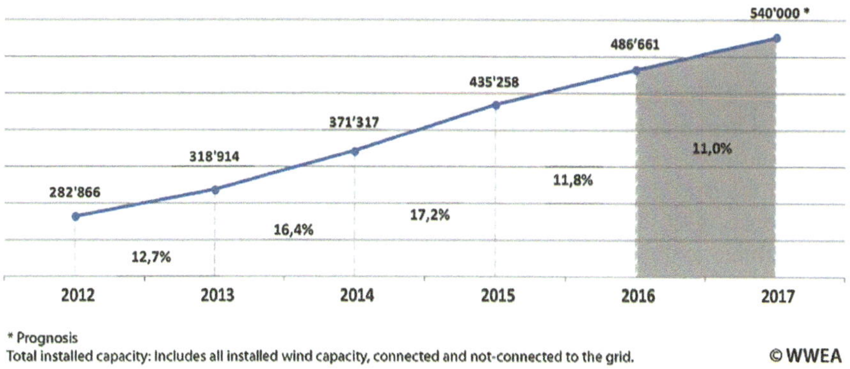

Fig. 2.6 Total world wind power installed capacity 2012–2016. *Source* World Wind Energy Association (WWEA) (2017)

Table 2.2 Total installed wind power in different countries in 2007

Country	Total capacity (MW)	Installed capacity in 2015 (MW)
China	148,000	32,970
United States	74,347	8,598
Germany	45,192	4,919
India	24,759	2,294
Spain	22,987	0

Source WEC (2016b)

these investments were made after the year 2000. The four countries that account for the bulk of these investment are: The United States, China, Germany and Spain. China has once more reaffirmed its role as the global wind power leader, adding 33 GW of new capacity, which represents a market share of 51.8% (WEC 2016b).

On the other hand, as reported by WEC (2016b), the US market also had a good performance the strongest growth since 2012, with 8.6 GW of added capacity. Germany, in turn, installed 4.9 GW in anticipation of changes in legislation. Brazil represented the fourth largest market for new turbines, with a market volume of 2.8 GW. By November 2015, India had installed around 2.3 GW of new capacity (WEC 2016b).

Regarding global wind power generation, the production amounted to 950 TWh in 2015, which represented roughly 4% of total global power generation. However, some countries reached much higher percentages on their national energy grid. For instance, in 2015, Denmark produced 42% of its electricity from wind turbines, the highest figure yet recorded worldwide, and Germany reached a new record of 13% of its power demand coming from wind power (WEC 2016b).

2.2.2 *Wind Energy Potential and Generation in Brazil*

According to the Brazilian Wind Energy Association (ABEEólica), wind energy utilization began in Brazil in 1992, with the commercial operation of the first wind turbine installed on Brazilian soil. This project was the result of a partnership between the Brazilian Wind Energy Center (CBEE) and Companhia Energética de Pernambuco (CELPE), funded by the Danish research institute Folkecenter. This wind turbine, with power capacity of 225 kW, installed in Fernando de Noronha (PE), was the first to enter commercial operation in South America. However, in the following decade, the advances in the consolidation of wind energy as a power generation alternative in the country were not very significant, due to factors discussed earlier, such as the lack of incentive policies and especially the high investment cost of this technology.

However, with the energy crisis in Brazil in 2001, the Brazilian government sought to promote business endeavors to generate wind power. Therefore, the Emergency Wind Energy Program—PROEÓLICA was created, aimed at contracting 1050 MW of wind energy projects by December 2003. In addition, the seasonal complementarity of the wind regime with hydrological flows in the hydroelectric reservoirs was already being discussed (Brazilian Eolic Energy Association (ABEEÓLICA) 2016).

However, the program was not successful and was replaced by the Alternative Energy Sources Incentive Program, PROINFA, as already mentioned. As a result of PROINFA, in addition to energy auctions and the development of the free market in 2012, Brazil had 108 wind farms totaling 2.5 GW of installed capacity.

According to the ranking published by the Global Wind Energy Council (GWEC), an international wind energy organization, in 2016, the country presented an expansion of 2014 MW in electricity generation. This advance placed Brazil in 5th place in the world ranking of installed capacity that same year. The country also occupied the new position in the world ranking of accumulated wind power capacity, reaching ninth place with a cumulative power of 10,740 MW (Abeeólica 2016).

Wind capacity in Brazil ranks among the largest wind power potential in the world. The Global Wind Energy Council (2015) points out that Brazil has wind power three times higher than the country's electricity needs, with Brazilian capacity factors above the world average. While the other countries have an average of 25%, Brazil surpasses the 50% capacity factor, reaching factors of more than 70% in times of the best winds. Thus, the quality of Brazilian wind and the capacity factor registered annually places the country in a prominent position in the world scenario with regard to generating wind power (Abeeólica 2016).

However, this capacity is not uniformly distributed throughout the national territory. According to Fig. 2.7, the Northeast is the Brazilian region that most stands out in relation to the wind potential, followed by the Southeast and South Regions. Regarding installed wind capacity, the Northeast Region also stands out, with Rio Grande do Norte in the lead, with 2243 MW installed, followed by Ceará (1233 MW), Rio Grande do Sul (1300 MW) and Bahia (959 MW).

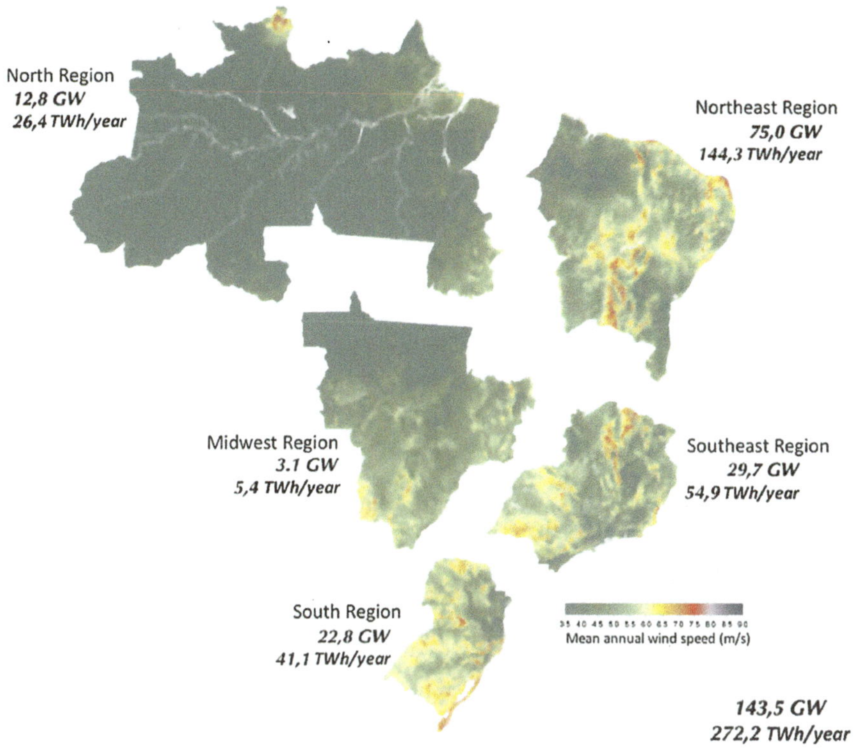

Fig. 2.7 Estimated wind potential per year for average wind speed equal to or greater than 7.0 m/s. *Source* Amarante et al. (2001)

2.2.3 Basic Principles of Wind Power Generation

Winds are formed due to pressure gradients caused by thermal gradients that result from the atmosphere's heating and cooling. These wind currents are not constant over time. According to Dunlap (2015), although average wind speed can be estimated at a particular location, its variations are somewhat more unpredictable. Thus, the use of wind power must be planned together with storage systems.

The kinetic energy in the wind currents can be transformed into electrical energy from a wind power generation system, Fig. 2.8 illustrates the basic components. This system consists of a windmill, or wind turbine, connected to a transformer and transmission system.

The wind pressure rotates the blades coupled to the rotor, which is connected to the spindle of an electric generator by means of transmission gear box. As a result, the kinetic energy of the wind is transformed into electric energy by the generator. This energy is produced in direct-current power. Thus, the invention of the "synchronous inverter" has revolutionized the wind energy industry as it has allowed the conversion of direct-current from the wind generator into alternating current, thus

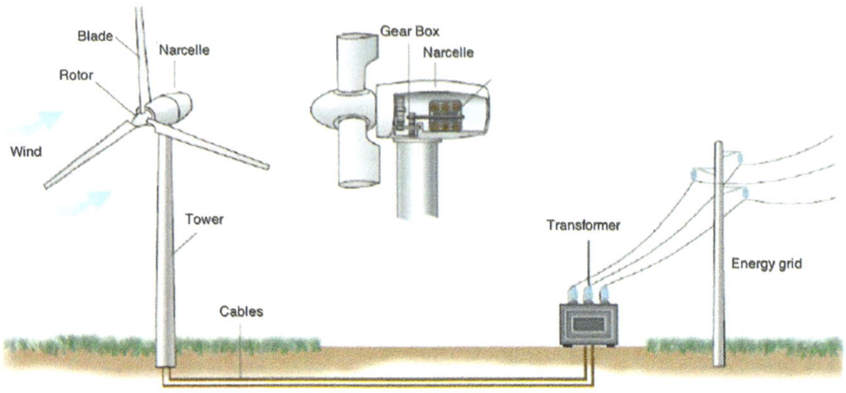

Fig. 2.8 Wind generation system. *Source* Adapted from Accettabilitasociale (2017)

enabling the connection for operation in parallel with the utility grid at the correct frequency (Hinrichs et al. 2010).

Modern wind turbines can be used to generate electric power in both isolated and grid-connected applications. These turbines are classified according to the orientation of the rotor axis with respect to the ground, and can be a horizontal axis or vertical axis. The conversion efficiency of each turbine depends on its configuration, and the two- or three-blade propeller is considered the most efficient for generating electricity because of its size and also because it is the most common one (Hinrichs et al. 2010). However, despite all its efficiency, not even this model is able to extract all the energy existing in the wind, because a turbine is incapable of presenting efficiency greater than Betz's Coefficient (59.3%). This coefficient refers to the wind speed at the output of the wind rotor which cannot be less than 1/3 of the incident wind speed. In this way, the rotor absorbs energy equivalent to 2/3 of the available energy in the free wind before the turbine (Custódio 2013).

In addition to its main components, the wind turbine, depending on its application, requires additional components to condition the power generated for the direct service of the loads or connection to the electric grid, such as cables, switches, transformers, batteries, among others. As the winds are intermittent; storing power for later use is very important if it is not possible to connect it to the existing power grid.

2.2.4 *Wind Energy Sustainability*

Wind energy has great potential to reduce greenhouse gas emissions by replacing electricity from fossil fuels (IPCC 2012). In the Brazilian context, while small wind farms are able to supply small localities distant from the grid, contributing to the universalization of energy services, large power plants can service a significant portion of the National Interconnected System (ANEEL 2008). Thus, the share of this energy source in the Brazilian matrix can bring significant environmental and social benefits, such as reduce the emission of atmospheric pollutants; reduce the need to build large reservoirs; and reduce the risks of energy shortages due to hydrological seasonality, thanks to the complementarity of the two systems. In addition, the commercial availability and the technology cost renders this type of utilization to be immediately implemented on a large scale (IPCC 2012).

However, wind power production can also generate negative social and environmental impacts. The main potential ecological impacts generated by onshore wind farms are related to the collision and death of birds and bats in addition to the indirect modifications of the habitat and ecosystem. These impacts can also be seen in offshore wind farms, coupled with implications for benthic resources, fisheries and marine life in general.

The collision with wind turbines can cause bird and bat deaths, which are among the best known environmental concerns associated with wind power plants. Avian fatality rates are particularly related to different types of plants and different types of bird species, and may vary according to region, site characteristics, climate, size, height and turbine design, among other factors (IPCC 2012). However, collision-related bird deaths in wind farms are at a lower scale when compared to other anthropogenic causes, such as: collisions with vehicles, buildings and windows, transmission lines, and communications towers; predation by domestic cats; and environmental pollution (Erickson et al. 2005).

In addition to the collisions, the possible effect of wind power plants on the local climate has also been the focus of research and discussions, although this remains vague due to the current insufficient amount of information and literature on this phenomenon. This change is due to the fact that the wind turbines extract the momentum of the airflow, reducing the wind speed behind the turbines, as well as increasing the vertical mix by introducing turbulence at several length scales (Baidya Roy and Traiteur 2010). Other studies point to these local effects that can affect the microclimate of the region, changing rainfall, radiation, clouds, wind direction and other climatic variables. While the overall average impact of these local changes is not very significant, they may have implications for local ecosystems and human activities (IPCC 2012).

Wind energy production may also be responsible for social impacts, especially with regard to visual and sound impacts. The former is the result of rotor noise and may vary according to the equipment. Thus, to avoid disruption to the neighboring population, it is important that the level of noise pollution emitted by the turbines complies with current legislation. The visual impacts are attributed to the grouping of towers and wind generators, mainly in the case of large wind farms.

As already mentioned, the tendency of the wind energy share in the national electricity matrix is that it will grow in the coming years, causing the existing concerns and impacts to worsen and new concerns to emerge. However, regardless of the type and degree of social and environmental impacts, it is imperative to directly address these concerns for any successful planning of new wind farms. For this to happen, it is essential to involve the local community in this planning, they should be actively present in the decision-making process and able to give opinions, for example, about alternative sites. According to Wolsink (2000), studies have shown that public reaction to wind farms has improved significantly when the planning and implementation process carried out is perceived as transparent.

2.3 Photovoltaic Solar Energy

As the name itself implies, solar energy uses the incoming solar radiation absorbed by Earth's atmosphere. This energy can be converted into heat, and used in heating systems, or converted into electricity by direct or indirect methods. While photovoltaic modules directly transform solar energy into electrical energy, concentration systems capture and concentrate the heat of the sun, using it to heat a fluid, which in turn is used to produce steam, responsible for powering turbines, whose mechanical energy is finally converted into electricity through generators.

Whether in the form of heat or electricity, the history of mankind's harnessing of the sun dates back to the intuitive discovery of the concepts of thermal comfort when early civilizations discovered that homes with openings to sunrise became warmer and brighter, even when the weather was cold. However, it was only in the second half of the 19th century that solar collectors were invented for transforming the energy from the sun into heat, which were applied in practical domestic use and in industrial applications. In the following decades, mirrors were used to raise the temperature of fluids, so that the heat engines powered by the energy of the sun could develop the driving force, and consequently electricity potential. The invention of the "photovoltaic cell" was in the late nineteenth century. This device was capable of converting sunlight directly into electricity, not requiring a heat engine. The modern model of silicon solar cells was discovered decades later in the 1940s by Russell Ohl while working at the American Telephone and Telegraph (AT&T) Bell Labs (IPCC 2012).

In the 1950s, the founding of the International Solar Energy Society (ISES) and increased research and development efforts in many industries gave rise to the modern era of solar research. Consequently, when the oil crisis began in the 1970s, several countries were increasing research and development programs for solar energy. These efforts have continued until now, making solar energy one of the fastest growing renewable technologies, with great potential to play a much larger role in the world energy matrix.

2.3.1 World Photovoltaic Power Generation

Solar energy is the only infinite renewable source that has the capacity to provide enough energy to meet the entire world energy demand. The total solar incidence in our atmosphere is about 1.8×10^{17} W. Of this total, approximately half is transmitted through the atmosphere, resulting in a sunshine of 9.0×10^{16}W on the earth's surface. Considering a photovoltaic panel with a modest efficiency of 15%, the resulting potential that can be supplied worldwide by these modules is 1.3×10^{16}W. Knowing that the total primary energy use in the world is about 5.7×10^{20} J, for a power demand of 1.8×10^{13} W, the use of only 0.14% of the available solar energy would be able to satisfy all our energy needs (Dunlap 2015).

Currently, the participation of solar energy is still not very expressive in the world energy grid. However, it increased by 2000% between 1996 and 2006, with a global installed capacity reaching 177 GW by 2015. The evolution of World installed solar photovoltaic capacity by country, from 2006 to 2012, and projected total installed capacity in 2020 (gigawatts) is represented in Fig. 2.9. Table 2.3 shows the relative participation of the countries in this production. China ranks as the world's largest producer, with 22.5% of the world total installed capacity. Together, China, Germany, Japan, the United States, and Italy account for 84% of world installed capacity (IEA 2016a), in 2015. All these are countries with strong programs of diversification and simultaneous "cleaning" of the local energy grid.

Regarding Annual Installed Capacity, in 2015, China, Japan and the United States reached the Top 3 countries in the world ranking, as represented in Table 2.4. According to IEA (2016a), while China's final installation numbers for 2013 and

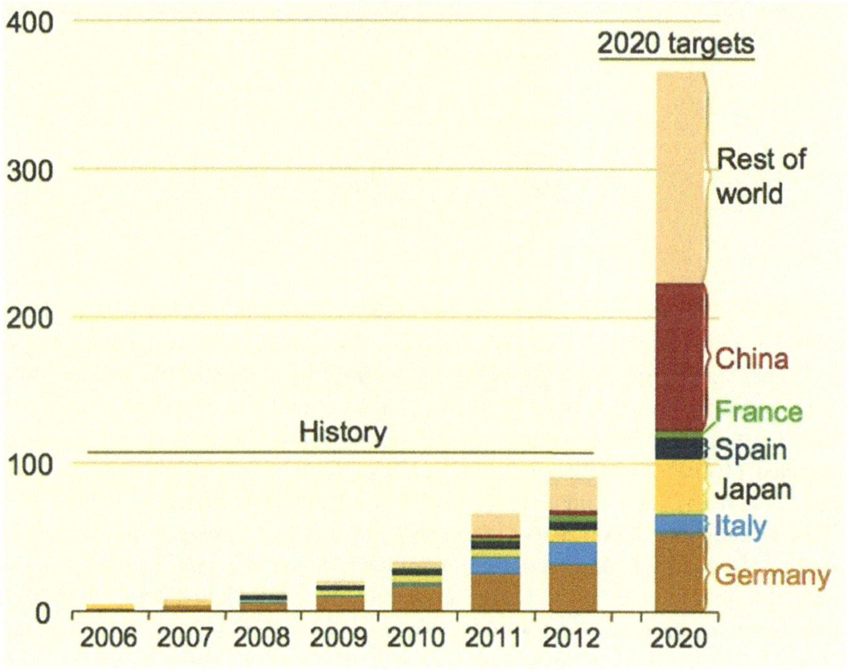

Fig. 2.9 World installed solar photovoltaic capacity by country, from 2006 to 2012, and projected total installed capacity in 2020 GW. *Source* IEA (2016b)

Table 2.3 Top 10 countries for total installed capacity in 2015

Country	Cumulative installed capacity in 2015 (MW)	% of world total installed capacity
China	43.53	22.5
Germany	39.7	20.6
Japan	34.41	17.8
United States	25.62	13.3
Italy	18.92	9.8
United Kingdom	8.78	4.5
France	6.58	3.4
Spain	5.4	2.8
Australia	5.07	2.6
India	5.05	2.6

Source IEA (2016a)

Table 2.4 Top 10 countries for PV installations in 2015

Country	Annual installed capacity in 2015 (MW)
China	15.3
Japan	11.0
United States	7.3
United Kingdom	3.5
India	2.0
Germany	1.5
Korea	1.0
Australia	0.9
France	0.9
Canada	0.6

Source IEA (2016a)

2014 reached respectively 10.95 and 10.6 GW, in 2015, this number grew significantly, reaching a total of 15.3 GW installed. This growth reflects the ambitions of the Chinese authorities on continuing to develop their internal PV market. Now China is aiming to reach a total of 143 GW of cumulative installed capacity by 2020. On the other hand, Japan was the second market for PV in 2015 with an estimated 11 GW of installations, and the USA was the third market, again in 2015, with 7.3 GW installed compared to 6.2 GW in 2014 (IEA 2016a).

As reported by IEA (2016a), together, the top five countries in the annual installed PV energy ranking represent more than 78% of the annual installations in 2015. Taking first place in the European market, the UK installed 3.5 GW in 2015, while Germany decreased to 1.46 GW, down from 1.9 GW in 2014. India installed around 2 GW in 2015, a relevant advance that may lead to a larger development in the next years. On the other hand, the top seven, represented by countries with more than 1 GW of annual installations, contributed to 83% of the global PV market.

However, several PV markets that had been developing in recent years declined or disappeared in 2015 (IEA 2016a). For instance, South Africa (200 MW est.) and Thailand (121 MW) are two of the countries of growing markets that have decreased significantly, although they remain promising for the future. Moreover, in Europe, some countries that were growing quickly have now stalled or experienced limited or greatly reduced additions, like Italy, that installed only 300 MW of PV systems in 2015, compared to 9.3 GW in 2011 (IEA 2016a). This reduction on PV market growth can be explained by the phase-out of the feed-in tariffs that were granted for new PV installations, which left the market driven by the self-consumption scheme and additional tax rebates. In addition, net-metering systems also collapsed once the regulations were changed. These systems were responsible for developing the PV market in several European countries (IEA 2016a).

Also according to IEA (2016a), South America, on its turn, has validated several GW of PV plants in Chile and a promising real development has materialized at last in the country with close to 450 MW in 2014 and 2015. In addition, countries like Mexico, Brazil and Peru, where the real PV market remained symbolic, have begun to adopt policies that promote the development of PV in the coming years. In the Middle East, the installation in Israel declined slightly, with around 200 MW installed in 2015, while installations in Turkey have started off with 208 MW in 2015. In Africa, South Africa declined after a really promising year in 2014, with around 200 MW installed, and Algeria has installed 268 MW in 2015 and has plans to continue the development of this market in the future.

2.3.2 *Photovoltaic Power Potential and Generation in Brazil*

Brazil has the most privileged areas in terms of high incidence rates of solar radiation, compared to countries that have already consolidated their participation in the solar energy market, according to Fig. 2.10, which shows the annual and daily average global solar radiation of each Brazilian region. In addition to being a country where the sun appears on average 280 days a year, Brazil holds one of the largest silicon reserves in the world, material used in the manufacture of solar panels. However, the use of all this energy potential still requires technological capabilities for the production of materials and solar systems by industries and there should be more incentives for this energy to become more competitive in the market (Cabral et al. 2013).

The less sunny region of Brazil has solar irradiation rates of around 1642 kWh/m^2, which are higher than the values in the highest solar incidence area in Germany, which receives about 1300 kWh/m^2 (Salamoni and Rüther 2007). However, in spite of the better climatic conditions, Brazil is far behind the European country, which has installed photovoltaic power above 35 GW, in contrast to the 23 MW installed in Brazil.

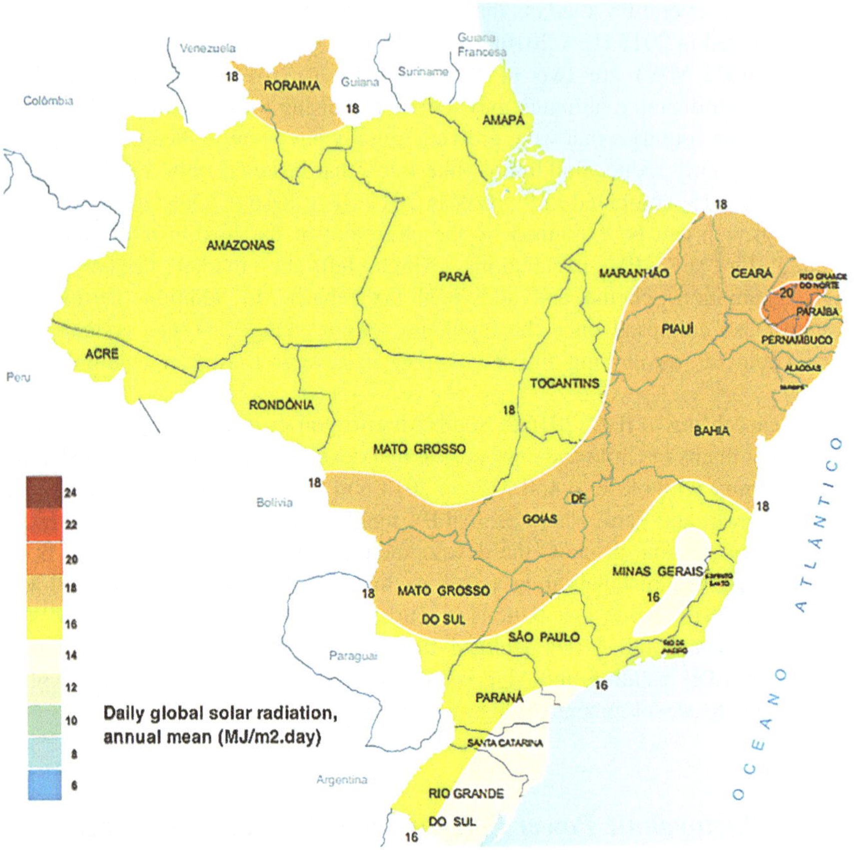

Fig. 2.10 Solarimetric atlas of Brazil. *Source* Tiba et al. (2000)

Until the recent past photovoltaic energy was used in Brazil only in isolated or autonomous systems installed in locations not reached by the electricity grid, such as rural properties, isolated communities, remote telecommunication centers and signaling systems. However, the approval of ANEEL's use of generation systems connected to the grid, through resolution 482 of 2012, allowed the recent advance of the number of photovoltaic systems connected to the grid, and a significant increase of these systems in the country is expected for the incoming years (Villalva 2015).

However, the consolidation of a significant share of solar energy in the national grid still has to overcome certain challenges. There was little initiative to boost the insertion of photovoltaic energy in the country until 2011, consequently, the share of this energy source in our system is practically negligible. In addition, until 2012, the absence of regulations and standards for the photovoltaic sector inhibited the emergence of industries and markets for distributed generation systems operating in low voltage (Villalva 2015).

In addition to the lack of necessary regulation, the absence of government incentives is another obstacle to the development of photovoltaic systems in Brazil, which requires programs that can encourage small producers, ordinary people or small businesses to acquire small generation systems. In addition, the cost of photovoltaic electricity is still considerably high compared to that of hydroelectric power, which can be seen as a major disadvantage for consumers.

2.3.3 Basic Principles of Photovoltaic Power Generation

The use of solar energy can be carried out employing a photovoltaic panel, due to a phenomenon known as photoelectric effect. The energy is generated by sunlight when it directly strikes the panels of semiconductor materials (silicon). These materials contain smaller cells, which are arranged in two layers, one positive and one negative. When the sun's energy arrives, the semiconductor material causes the electrons to move between the two layers, with an amount of kinetic energy inversely proportional to the wavelength of the incident light, and generate a continuous electric current. However, much of the light that falls on a cell is lost before it can be converted into an electric current, and the conversion efficiencies of a solar panel vary from 10 to 15% and may reach 30% (Hinrichs et al. 2010). Figure 2.10 shows the operating scheme of a photovoltaic module.

The solar photovoltaic system is composed of three basic sub-systems: the generator subsystem, the power conditioning sub-system and, where necessary, a storage unit. While the first sub-system consists of photovoltaic panels in various arrangements, electrical cabling and support structures, the second can be composed of dc-cc converters, maximum power points (MPPT), inverters, load controllers, among other protection and control devices. The third includes the batteries and/or other forms of storage. These systems may be connected to the network or stand-alone (Pinho and Galdino 2014) (Fig. 2.11).

Most photovoltaic cells in use today are composed of monocrystalline silicon, although other materials, such as polymers are gaining market share. The making of a solar panel involves a silica (SiO_2) purification and refinement process, followed by melting and solidification for the perfect arrangement of the silicon atoms. One of the methods used for this process is the introduction of a crystalline silicon seed into a pure silicon melt, followed by its slow extraction, forming a cylindrical ingot. This ingot is then sliced into wafers, which are doped with phosphorus impurities (n-type layer) or boron (p-type layer) to form the p-n junction, responsible for the photoelectric effect mechanism (Hinrichs et al. 2010).

Also according to these authors, more advanced crystal production techniques can reduce the manufacturing costs of solar cells, which are still considerably high. In addition to the monocrystalline modules, there are several other types of photovoltaic cells, already in the market and with various efficiencies. Two of them are those of polycrystalline silicon and those of amorphous silicon, both represent less expensive production.

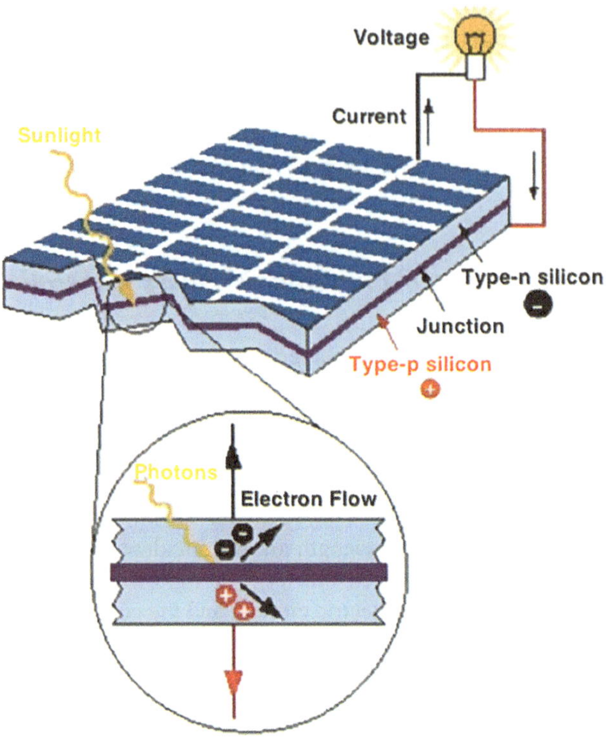

Fig. 2.11 Operation schematic of a photovoltaic panel. *Source* Energia Inteligente (2017)

2.3.4 Sustainability of Photovoltaic Energy

Photovoltaic systems are capable of servicing isolated communities disconnected from the national energy distribution system. Solar energy can usually be harnessed in places where it is impossible to explore other sources, such as hydroelectric or wind, due to the simplicity of the system and abundance of the light source. Therefore, this technology has the potential to improve the quality of life of these more isolated populations, generating new economic opportunities, better working conditions, and greater access to information.

Moreover, another advantage of a photovoltaic system is that it generates almost no solid, liquid or gaseous residue during operation, except for the water used in the sporadic cleaning of the panels. In addition, the operation of these systems does not make any noise and or require the use of any non-renewable energy source.

However, as discussed earlier, the production of energy from renewable sources is not entirely emission-free or without environmental impacts, and photovoltaic energy is no exception. Currently, there is no extensive literature on the impacts generated by this type of energy use, because it is a relatively new technology in the market. However, the main concerns raised relate to the generation of pollutants during the whole life cycle of a photovoltaic panel and to the possibility of recycling at the end of life of the generation systems.

Although no type of residue is significantly generated during the operation phase, the life cycle study of a photovoltaic module shows that during their production the emission of greenhouse gases and other toxins cannot be disregarded. Tables 2.1 and 2.2 show that, among the renewable energies studied, solar energy is the largest source of greenhouse gas emissions, NOx, and SO_2 when the entire solar 1 kWh life cycle is analyzed. These emissions are due to the production characteristics of the photovoltaic modules, which require an intensive concentration of materials such as Silicon and Cadmium Telluride. Moreover, considerable energy is consumed to extract these materials, necessary for mining an enormous amount of raw material to obtain a small quantity of the photovoltaic component.

Regarding the end of life of photovoltaic panels the recycling of modules is already economically viable, mainly for concentrated and large-scale applications. There are already reuse projections for some of the modules' materials, such as 80–96% of glass, ethylene vinyl acetate, and metals (ET, selenium and lead). However, other metals, such as Cd, ET, tin, nickel, aluminum and copper, must be stored or recycled by other methods (IPCC 2012).

In addition to the environmental impacts, social impacts related to the use of photovoltaic technology are also raised, because although this is a renewable source with great potential, it can cause resistance in the consumer public. One of the reasons for resistance is the aesthetic impact of a large solar plant, which would require large areas for its installation. However, this impact can be avoided by choosing more remote and less populated areas, which are ideal for implementing these plants.

With this in mind, it is necessary to devise strategies that seek to mitigate these social impacts, in order to reduce the possible resistance of communities to the installation of any type of renewable generation system, not only solar. Thus, it is important to integrate these communities into the system planning process, so they have more information and access to these alternative energy generation methods, and better understand both the importance, the advantages and disadvantages of each system, so that they have an active participation in the decision-making process.

References

Agência Nacional de Energia Elétrica (ANEEL) (2008) Atlas de energia elétrica do Brasil, 3rd edn. Brasília

Agência Nacional de Energia Elétrica (ANEEL) (2016b) Boletim de Informações Gerenciais 2016. Available at http://www.aneel.gov.br. Accessed 27 Feb 2017

Amarante OAC, Brower M, Zack J (2001) Atlas do Potencial Eólico Brasileiro. MME/ELETROBRAS/CEPEL

Associação Brasileira de Energia Eólica (ABEEÓLICA) (2016) Boletim Anual de Geração Eólica 2016. São Paulo. Available at http://www.portalabeeolica.org.br. Accessed 5 Aug 2017

Baidya Roy S, Traiteur JJ (2010) Impacts of wind farms on surface air temperatures. Proc Natl Acad Sci 107(42):17899–17904

Cabral IS, Torres AC, Senna PR (2013) Energia Solar – Análise Comparativa entre Brasil e Alemanha. IV Congresso Brasileiro de Gestão Ambiental, Salvador

Custódio RS (2013) Energia eólica para a produção de energia elétrica, 2nd edn. Synergia, Porto Alegre

Dunlap RA (2015) Sustainable energy. Dalhousie University, Halifax

Eletrobras (2015) Sistema de Informações do Potencial Hidrelétrico Brasileiro – SIPOT, Rio de Janeiro

Energia Inteligente. Como funciona: Energia Solar Fotovoltaica. Available at http://energiainteligenteufjf.com/como-funciona/energia-solar-fotovoltaica/. Accessed 11 Aug 2017

Erickson WP, Johnson GD, Young DP (2005) A Summary and comparison of bird mortality from anthropogenic causes with an emphasis on collisions. United States Forest Service, Washington (General Technical Report)

Global Wind Energy Council (GWEC) (2015) global wind market report. GWEC, Brussels. www.gwec.netAccessed 10 Aug 2017

Hinrichs RA, Bach M, Dos Reis LB (2010) Energia e meio ambiente. Cengage Learning, São Paulo

International Energy Agency (IEA) (2016a) 2015 A snapshot of global photovoltaic markets (Photovoltaic Power Systems Program—Report IEA PVPS T1-29:2016). http://www.iea-pvps.org/fileadmin/dam/public/report/PICS/IEA-PVPS_-__A_Snapshot_of_Global_PV_-_1992-2015_-_Final_2_02.pdf Acessed: 23 June 2016

International Energy Agency (IEA) (2016b) International energy outlook 2016. U.S. Energy Information Administration, Washington. Available at www.eia.gov/forecasts/ieo/pdf/0484(2016).pdf . Accessed 3 Aug 2017

Intergovernmental Panel on Climate Change (IPCC) (2012) Renewable energy sources and climate change mitigation: special Report of the Intergovernmental Panel on Climate Change. Cambridge University Press, New York

Mendes NAS (2005) As usinas hidrelétricas e seus impactos: os aspectos socioambientais e econômicos do Reassentamento Rural de Rosana - Euclides da Cunha Paulista. UNESP, Presidente Prudente

Pinho JT, Galdino MA (2014) Manual de engenharia para sistemas fotovoltaicos. Ed. rev. e atual. Rio de Janeiro: Grupo de Trabalho de Energia Solar – GTES, CEPEL,CRESESB

Rogner HH, Barthel F, Cabrera M, Faaij A, Giroux M, Hall D, Kagramanian V, Kononov S, Lefevre T, Nötstaller R, Odell P, Taylor M (2000) Energy resources. In: Goldemberg J (ed) World Energy Assessment, United Nations Development Program, New York

Salamoni I, Rüther R (2007) Potencial Brasileiro da Geração Solar Fotovoltaica conectada à Rede Elétrica: Análise de Paridade de Rede. IX Encontro Nacional e V Latino Americano de Conforto no Ambiente Construído, Ouro Preto

Sousa WL (2000) Impacto ambiental de hidrelétricas: Uma análise comparativa de duas abordagens. Dissertação Mestrado, UFRJ, Rio de Janeiro

Tiba C, et al (2000) Atlas solarimétrico do Brasil: banco de dados terrestres. Ed. Universitária da UFPE, Recife

Villalva MG (2015) Energia solar fotovoltaica: conceitos e aplicações, 3 edn. Érica, São Paulo

World Energy Council (WEC) (2016a) World energy resources 2016. London

World Energy Council (WEC) (2016b) World Energy resources wind 2016. London

World Energy Council (WEC) (2016c) World energy resources hydropower 2016. London

World Energy Council (WEC). Data. Available at https://www.worldenergy.org/data/resources/resource/hydropower/. Accessed 21 Nov 2017

Worldwide Fund for Nature (2012) Sumário para tomadores de decisão: Além de Grandes Hidrelétricas. Políticas para fontes renováveis de energia elétrica no Brasil. Brasilia. Available at http://d3nehc6yl9qzo4.cloudfront.net/downloads/alem_de_grandes_hidreletricas_sumario_para_tomadores_de_decisao.pdf. Accessed 24 Mar 2017

World Wind Energy Association (WWEA) (2017) World Wind Market has reached 486 GW from where 54 GW has been installed last year. Available at http://www.wwindea.org/. Accessed 21 Nov 2017

Wolsink M (2000) Wind power and the NIMBY-myth: institutional capacity and the limited significance of public support. Renewable Energy 21:49–64

Chapter 3
Sustainabilty Indicators

Abstract An initial discussion on the triple bottom line is presented followed by a description on the main tendencies on its definition and uses. It compiles and discuss the indicators from the United Nations Commission on Sustainable Development (CDS). A criterion for the selection of sustainability indicators is presented with focus on energy sustainability. Economic, environmental and social indicators are then shown with regard to the International Atomic Energy Agency (IAEA) data with contributions from UNDESA, the International Energy Agency (IEA) and other international and national organizations.

Sustainable development can be defined as a growth based on the balance between its economic, environmental and social dimensions. In this way, this development should meet the needs of the present without compromising the ability of future generations to fulfill their own needs. Since Rio-92, the world has experimented, on different scales, major advances towards the promotion of sustainable development, which is increasingly present in discussions and in the main planning objectives of decision-makers. However, the application of sustainable development requires constant construction, evaluation and readjustment of objectives and indicators, so it should be possible to measure the evolution of a country or region in relation to what was established to achieve.

This chapter presents an overview on the definitions of sustainability and sustainable development, highlighting what are the relevant sustainability indicators and how to select them. Moreover, this chapter also introduces which of those indicators can be applied in the assessment of energy sustainability. This analysis is essential for the determination of which indicators are crucial for the methodology of sustainability performance evaluation of renewable energy sources, developed in Chaps. 4 and 5.

J. F. G. de Oliveira and T. C. G. Trindade, *Sustainability Performance Evaluation of Renewable Energy Sources: The Case of Brazil*,
https://doi.org/10.1007/978-3-319-77607-1_3

3.1 The Triple Bottom Line

The most accepted and well known definition of Sustainable Development resulted from the work of the Brundtland Commission. This work defines Sustainable Development as a development "that meets the needs of the present without compromising the ability of future generations to meet their own needs" (Brundtland Comission 1987). Thereafter, this concept of sustainability has been a crucial part of development work since the late 1980s. However, the understandings and the importance of sustainability have evolved significantly in the last 25 years since the Brundtland Commission. Thus, defining sustainability is not a simple task, given that the concept has now a broad and deep meaning that depends on many factors.

In the extensive discussion and application of the sustainability concept, since the Brundtland Commission, there has been an increasing recognition of three essential aspects of sustainable development, or **the triple bottom line**. A comprehensive definition is presented by Harris (2003):

- *Economic: An economically sustainable system must be able to produce goods and services on a continuing basis, to maintain manageable levels of government and external debt, and to avoid extreme sectoral imbalances which damage agricultural or industrial production.*
- *Environmental: An environmentally sustainable system must maintain a stable resource base, avoiding over-exploitation of renewable resource systems or environmental sink functions, and depleting nonrenewable resources only to the extent that investment is made in adequate substitutes. This includes maintenance of biodiversity, atmospheric stability, and other ecosystem functions not ordinarily classed as economic resources.*
- *Social: A socially sustainable system must achieve fairness in distribution and opportunity, adequate provision of social services including health and education, gender equity, and political accountability and participation.*

Another widely cited definition of environmental sustainability was supported by Goodland (1995). In his work, this concept is described as the effort society makes to *"improve human welfare by protecting the sources of raw materials used for human needs and ensuring that the sinks for human wastes are not exceeded, in order to prevent harm to humans"* (Goodland 1995). According to Goodland (1995), the need for sustainability came from the recognition that the current patterns of development lead to biophysical impossibilities when they are projected into the near future. In other words, our current pattern of development cannot be sustained by the Earth's resources for much longer. Therefore, the transition to sustainable development is imperative, for the reverting the deterioration of global life-support systems. As we don't have enough knowledge or time to colonize the moon or Mars, it is crucial that we save what is left of the only environment we have, providing with more efforts for the regeneration of what has been already damaged.

3.2 Sustainability Indicators

Sustainability indicators are tools to aid in the monitoring of the operationalization of sustainable development. Their main function is to provide information on the state of the critical dimensions that compose the concept of sustainability (Carvalho et al. 2011). Thus, these indicators are used to gather and quantify information, simplifying the understanding of complex phenomena, in order to improve the communication process (van Bellen 2006). The indicators should not only be used to serve the interests of the Public Power in assessing the efficiency and effectiveness of the policies adopted, but should also be used by citizens, as a citizenship instrument, as the indicators are able to provide valuable information on the state of the environment and quality of life (Santos and Câmara 2002).

In 1992, the United Nations Conference on Environment and Development recognized the importance of the role that indicators could play in aiding countries in their decision making concerning sustainable development. Internationally, in 1995, the Work Program on Indicators of Sustainable Development was approved by the Commission on Sustainable Development (CSD). Consequently, between 1994 and 2001, the first two sets of CSD Indicators of Sustainable Development (CSD indicators) were developed (United Nations 2007). After that, these sets have been extensively applied in several countries, basing the development of their own set of national indicators of sustainable development.

In 2002, the CSD and the World Summit on Sustainable Development encouraged additional work on indicators at the country level, aligned with national conditions and priorities, and invited the international community to support the efforts of developing countries (United Nations 2007). In response to these decisions, a new revised edition of the CSD indicators has been developed. However, since the publication of the first set, the experience of countries and organizations with sustainable development indicators has greatly increased.

According to the United Nations (2007), the newly revised CSD indicators are composed by a core set of 50 items, which are part of a larger set of 96 indicators of sustainable development. These core indicators must fulfill three different criteria:

- First: they must cover issues relevant for sustainable development in most countries;
- Second: they must provide critical information, not available from other core indicators;
- Third: they can be calculated by most countries with data that is readily available or could be reasonably made available, regarding cost and time.

It is important to notice that the indicator set keeps the same thematic/sub-thematic framework adopted in 2001, in order to remain consistent with the practice of most countries applying national sustainable development indicator sets (United Nations 2007; Goldemberg and Lucon 2007). In the new set, the indicators are no longer explicitly divided in four 'pillars' (social, economic, environmental and institutional), emphasizing the multi-dimensional nature of sustainable development and reflecting the importance of integrating all different dimensions of sustainability. In this new set, the main CDS indicator themes are: Poverty; Governance; Health; Education; Demographics; Natural hazards; Atmosphere; Land; Oceans, seas and coasts; Freshwater; Biodiversity; Economic development; Global economic partnership; Consumption and production patterns.

Tables 3.1, 3.2, 3.3, 3.4, 3.5, 3.6, 3.7, 3.8, 3.9, 3.10, 3.11, 3.12 and 3.13 compile all the proposed CDS indicators by each different theme. The core indicators are the ones that fulfill the three criteria mentioned before. On the other hand, the indicators that are not part of the core (Other indicators) are either important only for some countries, present complementary information to core indicators or are not quite available for most countries (United Nations 2007) (Table 3.14).

Table 3.1 CDS indicators: poverty theme

Theme	Sub-theme	Core indicator	Other indicator
Poverty	Income poverty	Proportion of population living below national poverty line	Proportion of population below $1 a day
	Income inequality	Ratio of share in national income of highest to lowest quintile	
	Sanitation	Proportion of population using an improved sanitation facility	
	Drinking water	Proportion of population using an improved water source	
	Access to energy	Share of households without electricity or other modern energy services	Percentage of population using solid fuels for cooking
	Living conditions	Proportion of urban population living in slums	

Source United Nations (2007)

Table 3.2 CDS indicators: governance theme

Theme	Sub-theme	Core indicator	Other indicator
Governance	Corruption	Percentage of population having paid bribes	
	Crime	Number of intentional homicides per 100,000 population	

Source United Nations (2007)

Table 3.3 CDS indicators: health theme

Theme	Sub-theme	Core indicator	Other indicator
Health	Mortality	Under-five mortality rate Life expectancy at birth	Healthy life expectancy at birth
	Health care delivery	Percent of population with access to primary health care facilities Immunization against infectious childhood diseases	Contraceptive prevalence rate
	Nutritional status	Nutritional status of children	
	Health status and risks	Morbidity of major diseases such as HIV/AIDS, malaria, tuberculosis	Prevalence of tobacco use Suicide rate

Source United Nations (2007)

Table 3.4 CDS indicators: demographics theme

Theme	Sub-theme	Core indicator	Other indicator
Demographics	Population	Population growth rate Dependency ratio	Total fertility rate
	Tourism		Ratio of local residents to tourists in major tourist regions and destinations

Source United Nations (2007)

Table 3.5 CDS indicators: education theme

Theme	Sub-theme	Core indicator	Other indicator
Education	Education level	Gross intake ratio to last grade of primary education Net enrolment rate in primary education Adult secondary (tertiary) schooling attainment level	Lifelong learning
	Literacy	Adult literacy rate	

Source United Nations (2007)

Table 3.6 CDS indicators: atmosphere theme

Theme	Sub-theme	Core indicator	Other indicator
Atmosphere	Climate change	Carbon dioxide emissions	Emissions of greenhouse gases
	Ozone layer depletion	Consumption of ozone depleting substances	
	Air quality	Ambient concentration of air pollutants in urban areas	

Source United Nations (2007)

Table 3.7 CDS indicators: oceans, seas and coasts theme

Theme	Sub-theme	Core indicator	Other indicator
Oceans, seas and coasts	Coastal zone	Percentage of total population living in coastal areas	Bathing water quality
	Fisheries	Proportion of fish stocks within safe biological limits	
	Marine environment	Proportion of marine area protected	Marine trophic index Area of coral reef ecosystems and percentage live cover

Source United Nations (2007)

Table 3.8 CDS indicators: economic development theme

Theme	Sub-theme	Core indicator	Other indicator
Economic development	Macroeconomic performance	Gross domestic product (GDP) per capita Investment share in GDP	Gross saving Adjusted net savings as percentage of gross national income (GNI) Inflation rate
	Sustainable public finance	Debt to GNI ratio	
	Employment	Employment-population ratio Labor productivity and unit labor costs Share of women in wage employment in the non-agricultural sector	Vulnerable employment
	Information and communication technologies	Internet users per 100 population	Fixed telephone lines per 100 population Mobile cellular telephone subscribers per 100 population
	Research and development		Gross domestic expenditure on R&D as a percent of GDP
	Tourism	Tourism contribution to GDP	

Source United Nations (2007)

Table 3.9 CDS indicators: fresh water theme

Theme	Sub-theme	Core indicator	Other indicator
Fresh water	Water quantity	Proportion of total water resources used Water use intensity by economic activity	Total fertility rate
	Water quality	Presence of faecal coliforms in freshwater	Biochemical oxygen demand in water bodies Wastewater treatment

Source United Nations (2007)

Table 3.10 CDS indicators: land theme

Theme	Sub-theme	Core indicator	Other indicator
Land	Land use and status		Land use change Land degradation
	Desertification		Land affected by desertification
	Agriculture	Arable and permanent cropland area	Fertilizer use efficiency Use of agricultural pesticides Area under organic farming
	Forests	Proportion of land area covered by forests	Percent of forest trees damaged by defoliation Area of forest under sustainable forest management

Source United Nations (2007)

Table 3.11 CDS indicators: biodiversity theme

Theme	Sub-theme	Core indicator	Other indicator
Biodiversity	Ecosystem	Proportion of terrestrial area protected, total and by ecological region	Management effectiveness of protected areas Area of selected key ecosystems Fragmentation of habitats
	Species	Change in threat status of species	Abundance of selected key species Abundance of invasive alien species

Source United Nations (2007)

Table 3.12 CDS indicators: economic development theme

Theme	Sub-theme	Core indicator	Other indicator
Economic development	Macroeconomic performance	Gross domestic product (GDP) per capita Investment share in GDP	Gross saving Adjusted net savings as percentage of gross national income (GNI) Inflation rate
	Sustainable public finance	Debt to GNI ratio	
	Employment	Employment-population ratio Labor productivity and unit labor costs Share of women in wage employment in the non-agricultural sector	Vulnerable employment
	Information and communication technologies	Internet users per 100 population	Fixed telephone lines per 100 population Mobile cellular telephone subscribers per 100 population
	Research and development		Gross domestic expenditure on R&D as a percent of GDP
	Tourism	Tourism contribution to GDP	

Source United Nations (2007)

Table 3.13 CDS indicators: global economic partnership theme

Theme	Sub-theme	Core indicator	Other indicator
Global Economic partnership	Trade	Current account deficit as percentage of GDP	Share of imports from developing countries and from LDCs Average tariff barriers imposed on exports from developing countries and LDCs
	External financing	Net official development assistance (ODA) given or received as a percentage of GNI	Foreign direct investment (FDI) net inflows and net outflows as percentage of GDP Remittances as percentage of GNI

Source United Nations (2007)

Table 3.14 CDS indicators: consumption and production patterns theme

Theme	Sub-theme	Core indicator	Other indicator
Consumption and production patterns	Material consumption	Material intensity of the economy	Domestic material consumption
	Energy use	Annual energy consumption, total and by main user category Intensity of energy use, total and by economic activity	Share of renewable energy sources in total energy use
	Waste generation and management	Generation of hazardous waste Waste treatment and disposal	Generation of waste Management of radioactive waste
	Transportation	Modal split of passenger transportation	Modal split of freight transport Energy intensity of transport

Source United Nations (2007)

3.3 Selecting Sustainability Indicators

According to Afgan et al. (1999), in order to deal with the complexity of the sustainability problems related to different systems, the indicators must reflect both the totality of the system and the interaction of its subsystems. Consequently, indicators should measure the intensity of interactions between elements of the system and between the system and its environment. In this view, the indicators selection should take into consideration the need for them to allow an evaluation of the complex relationship of each system and its environment. Thus, an effective indicator should address characteristics that reflect the problem and the criteria to be considered. The main purpose of these indicators should be to show how well the system is working, making it strongly dependent on the type of system that is being monitored (Afgan et al. 1999).

Aside from its purpose, there are other important criteria that should be taken into consideration for selecting a sustainability indicator. According to the United Nations (2007), the CSD indicator guidelines and methodologies recommend that sustainability indicators should be:

1. Primarily national in scope;
2. Important to assessing sustainable development progress;
3. Limited in number, remaining open-ended and adaptable to future demands;
4. Broad in coverage of Agenda 21 and all aspects of sustainable development;
5. Comprehensible, clear and unambiguous;
6. Conceptually sound;

7. Representative of an international consensus to the extent possible;
8. Within the feasibility of national governments to develop;
9. Dependent on cost effective data of known quality.

In other words, the selected indicators must be both limited and sufficiently understandable to seize the multidimensional nature of sustainable development (United Nations 2007). If a too extensive number of indicators is used, the results may become difficult to interpret and manage. In addition, the selected indicators should also be clear and unambiguous. This clarity can be achieved by the clear comprehension of the indicator's purpose and the audience who will interpret it. Indicators must also always be conceptually sound and within the national capability to develop it, which refers to the institutional capacity not only to collect data and monitor the indicator but also to interpret and synthesize it into useful information for decision-making (United Nations 2007).

It is important to point out that one major challenge for the indicator selection and application is the limited data availability and its reliability, which continues to be a problem in many countries. According to the United Nations (2007), to increase cost effectiveness, the indicators should often request information that are routinely collected either by national statistical services or through international processes.

3.4 Energy Sustainability Indicators

In order to provide accessible information on current energy-related trends to aid decision making at the national level, the International Atomic Energy Agency (IAEA), in cooperation with the United Nations Department of Economic and Social Affairs (UNDESA), the International Energy Agency (IEA), Eurostat and the European Environment Agency (EEA), developed a set of Energy Indicators for Sustainable Development (EISD) and corresponding methodologies and guidelines. These indicators were elaborated in the Johannesburg Plan of Implementation (JPOI) and their main purpose is to help countries with the assessment of effective energy policies for action on sustainable development (International Atomic Energy Agency (IAEA) 2005).

The EISD are able to assist the implementation of the actions urged at the World Summit on Sustainable Development (WSSD), held in Johannesburg. According to the IAEA (2005), these actions are:

1. Integrate energy into socioeconomic programmes;
2. Combine more renewable energy, energy efficiency and advanced energy technologies to meet the growing need for energy services
3. Increase the share of renewable energy options;
4. Reduce the flaring and venting of gas;
5. Establish domestic programmes on energy efficiency;

6. Improve the functioning and transparency of information in energy markets;
7. Reduce market distortions;
8. Assist developing countries in their domestic efforts to provide energy services to all sectors of their populations.

Although these indicators were developed by the intensive effort of all cited agencies, it is important to point out that no set of energy indicators can be final and definitive. Over time, these indicators have to evolve to be suitable to country-specific conditions, priorities and capabilities (IAEA 2005). With this in mind, the main goal of the established set of EISD is to serve as a starting point for the development of a more understandable and universally accepted set of sustainable energy indicators.

3.4.1 Background

Based on the effort of the United Nation to establish the CDS indicators, the initial work on indicators that were more specific for energy sustainability was undertaken by the International Atomic Energy Agency (IAEA) with contributions from UNDESA, the International Energy Agency (IEA) and other international and national organizations. The result of this work was presented at the ninth session of the Commission on Sustainable Development (CSD-9) in 2001, under the name 'Indicators for Sustainable Energy Development' (ISED).

The CSD-9 had energy as a major theme, focusing significantly on the improvement of affordability and accessibility to modern energy services and the promotion of less wasteful and more efficient use of energy resources. According to the IAEA (2005), the international community started to notice that relevant data could be able to guide decision makers in the development of suitable policy on energy supply options. Consequently, the energy indicators also began to be perceived as a tool for monitoring the consequences of these choices.

Back in 2002, the World Summit on Sustainable Development (WSSD) discussed the energy question. Based on the decisions taken at CSD-9, the international community reconfirmed 'access to energy' as relevant for the achievement of the Millennium Development Goals. In addition, the Summit also demanded changes in the current unsustainable patterns of energy production and use. As a result of this Summit, the Johannesburg Plan of Implementation (JPOI) was created, urging all nations, groups and institutions to take action to achieve the goals of sustainable development appointed in Agenda 21 and at the Earth Summit +5, and further (IAEA 2005).

Since energy sustainability had become undoubtedly relevant for the achievement of sustainable development, it urged for the establishment of indicators that were more specific to it, so energy sustainability could be better understood and measured in the National level by different countries. As a result, an interagency effort led by the IAEA in cooperation with UNDESA, the IEA, the Statistical Office of the European Communities (Eurostat) and the European Environment Agency (EEA), were responsible for the development of a set of energy sustainability indicators that will be discussed in more details in the following sections.

This set is composed by 30 indicators, classified into three different dimensions: social, economic and environmental. These are further classified into 7 themes and 19 sub-themes. Some of these indicators can be classified in more than one dimension, theme or sub-theme, considering the interlinkages among the categories (IAEA 2005). Moreover, each indicator may also represent a group of related indicators that are important for the evaluation of a particular issue. These IAEA indicators will be briefly described in the next 3 sections. They are included in this book in order to provide to government readers a clear overview on the mains aspects that should govern an energy development planning.

3.4.2 Economic Indicators

As a reliable and adequate energy supply is a prerequisite for industrialization and services, there is a significant demand on modern energy services by all sectors of the economy. Consequently, these services are responsible for promoting economic and social development at the local level, as they raise productivity and enable local income generation. Therefore, energy supply affects not only the number of offered jobs, but also productivity and development.

With this is mind, the IAEA (2005) divided the energy economic indicators in two different themes: Use and Production Patterns, and Security. The first theme has as sub-themes: Overall Use, Overall Productivity, Supply Efficiency, Production, End Use, Diversification (Fuel Mix) and Prices. The second has the sub-themes of Imports and Strategic Fuel Stocks. All these indicators are represented in Table 3.15, along with their major components.

While the 'energy use per unit of GDP' (ECO2) represents a marker of aggregate energy intensity, 'efficiency of energy conversion and distribution' (ECO3) monitors energy efficiency in transformation processes such as power stations. There are also indicators for energy intensity in individual sectors, which are sector specific, representing good benchmarks of energy efficiency, economic structure and the level of modernization of plants and equipment (IAEA 2005).

'Fuel shares in energy and electricity' (ECO11) presents the proportions of energy from different energy fuels, providing a picture of the primary energy supply mix and showing the extent of energy diversification. In its turn, the 'prices of end-use energy by fuel and sector' (ECO14) have great economic importance, since efficient energy pricing is key to efficient energy supply and use, and socially efficient levels of pollution abatement (IAEA 2005).

Regarding the second theme (Security), according to the IAEA (2005), one of the major objectives in the sustainable development criteria of many countries is providing energy security. Serious financial and economic losses can result of interruptions of energy supply. So, to support the goals of sustainable development,

Table 3.15 IAEA Economic Sustainable Energy Indicators

Economic				
Theme	Sub-theme	Energy indicator		Components
Use and production patterns	Overall use	ECO1	Energy use per capita	– Energy use (total primary energy supply, total final consumption and electricity use) – Total population
	Overall productivity	ECO2	Energy use per unit of GDP	– Energy use (total primary energy supply, total final consumption and electricity use) – GDP
	Supply efficiency	ECO3	Efficiency of energy conversion and distribution	– Losses in transformation systems including losses in electricity generation, transmission and distribution
	Production	ECO4	Reserves-to- production ratio	– Proven recoverable reserves – Total energy production
		ECO5	Resources-to-production ratio	– Total estimated resources – Total energy production
	End use	ECO6	Industrial energy intensities	– Energy use in industrial sector and by manufacturing branch – Corresponding value added
		ECO7	Agricultural energy intensities	– Energy use in agricultural sector – Corresponding value added
		ECO8	Service/commercial energy intensities	– Energy use in service/ commercial sector – Corresponding value added
		ECO9	Household energy intensities	– Energy use in households and by key end use – Number of households, floor area, persons per household, appliance ownership
		ECO10	Transport energy intensities	– Energy use in passenger travel and freight sectors and by mode – Passenger-km travel and tonne-km freight and by mode
	Diversification (fuel mix)	ECO11	Fuel shares in energy and electricity	– Primary energy supply and final consumption, electricity generation and generating capacity by fuel type

(continued)

Table 3.15 (continued)

Economic				
Theme	Sub-theme	Energy indicator		Components
				– Total primary energy supply, total final consumption, total electricity generation and total generating capacity
		ECO12	Non-carbon energy share in energy and electricity	– Primary supply, electricity generation and generating capacity by non-carbon energy – Total primary energy supply, total electricity generation and total generating capacity
		ECO13	Renewable energy share in energy and electricity	– Primary energy supply, final consumption and electricity generation and generating capacity by renewable energy
				– Total primary energy supply, total final consumption, total electricity generation and total generating capacity
	Prices	ECO14	End-use energy prices by fuel and by sector	– Energy prices (with and without tax/subsidy)
Security	Imports	ECO15	Net energy import dependency	– Energy imports – Total primary energy supply
	Strategic fuel stocks	ECO16	Stocks of critical fuels per corresponding fuel consumption	– Stocks of critical fuel (e.g. oil, gas, etc.) – Critical fuel consumption

Source IAEA (2005)

it is imperative that sufficient and affordable energy is available all of the time. Thus, secure energy supplies that provide reliable energy services to society are essential in order to maintain the economic activity.

3.4.3 Environmental Indicators

The exploitation of energy sources is bound to create environmental impacts. This happens in the production, distribution and use of energy, resulting in environment impacts in the household, workplace and city, and at the national, regional or global levels. These impacts are greatly dependent on how energy is produced and used, on the fuel mix, on the structure of the energy systems and on the related energy regulatory actions and pricing structures (IAEA 2005). As represented in Table 3.16, the suggested IAEA Environmental indicators are divided into three themes: Atmosphere, Water and Land.

Table 3.16 IAEA Environmental Sustainable Energy Indicators

Environmental				
Theme	Sub-theme	Energy indicator		Components
Atmosphere	Climate change	ENV1	GHG emissions from energy production and use per capita and per unit of GDP	– GHG emissions from energy production and use – Population and GDP
	Air quality	ENV2	Ambient concentrations of air pollutants in urban areas	– Concentrations of pollutants in air
		ENV3	Air pollutant emissions from energy systems	– Air pollutant emissions
Water	Water quality	ENV4	Contaminant discharges in liquid effluents from energy systems including oil discharges	– Contaminant discharges in liquid effluents
Land	Soil quality	ENV5	Soil area where acidification exceeds critical load	– Affected soil area – Critical load
	Forest	ENV6	Rate of deforestation attributed to energy use	– Forest area at two different times – Biomass utilization
	Solid waste generation and management	ENV7	Ratio of solid waste generation to units of energy produced	– Amount of solid waste
	Solid waste generation and management	ENV8	Ratio of solid waste properly disposed to total generated solid waste	– Energy produced – Total amount of solid waste
		ENV9	Ratio of solid radioactive waste to units of energy produced	– Amount of radioactive waste (cumulative for a selected period of time) – Energy produced
		ENV10	Ratio of solid radioactive waste awaiting disposal to total generated solid radioactive waste	– Amount of radioactive waste awaiting disposal – Total volume of radioactive waste

Source IAEA (2005)

The main issue addressed by the atmosphere theme include acidification, formation of tropospheric ozone and emissions of other pollutants affecting urban air quality. Aside from the Greenhouse gases emissions, another central point of discussion regarding anthropogenic environmental impacts are the air pollutants of major concern, also considered by the indicators, including Sulphur oxides, nitrogen oxides, carbon monoxide and particulates (IAEA 2005).

Other relevant sub-themes of the environmental dimension concern water and land quality. Land degradation and acidification might be caused by energy activities, affecting the quality of water and agricultural productivity. Moreover, the use of wood as fuel can result in deforestation, leading to erosion and soil loss.

According to the IAEA (2005), land can also be affected by energy transformation processes that often produce solid wastes, which include radioactive wastes, requiring adequate disposal. On the other hand, water quality may be affected by the discharge of contaminants in liquid effluents from energy systems, particularly from the mining of energy resources (IAEA 2005).

3.4.4 Social Indicators

Energy availability impacts employment, education, demographic transition, pollution and health. The IAEA (2005) established two main themes of regarding the social dimension: Equity and Health. Considering that social equity is one of the most important values underlying sustainable development, it is essential to assess the degree of fairness and inclusiveness with which energy resources are distributed, to ensure that energy systems are made accessible and pricing schemes are developed to ensure affordability.

With this in mind, the Equity indicators have the following sub-themes: accessibility, affordability and disparities. A lack or limited access to energy may lead to a disparity in the portion of income spent on energy between poor and rich households. Moreover, the lack of electricity also limits work opportunities and productivity, considering that today electricity is essential to use even the simplest tools and equipment. Regarding affordability, regions and income groups within a region may have different conditions to access or afford energy. The disparities within a country or between countries can be caused by uneven income distributions, inadequate energy distribution networks, and geographical differences among regions (IAEA 2005).

While the accessibility and affordability indicators are clear markers of progress towards development, the health indicators seek to eliminate the undesirable consequences that energy systems may cause on humans. Energy use might have negative impacts to human health, due to pollution or accidents. However, these problems should be as mitigated as possible, once the main objective of exploiting an energy source is to improve the living standards of the community which will receive it, consequently improving their health. Thus, a social goal is to reduce or eliminate these negative impacts. The Health indicators have the sub-theme of

Table 3.17 IAEA social sustainable energy indicators

Social				
Theme	Sub-theme	Energy Indicator		Components
Equity	Accessibility	SOC1	Share of households (or population) without electricity or commercial energy, or heavily dependent on non-commercial energy	– Households (or population) without electricity or commercial energy, or heavily dependent on non-commercial energy
				– Total number of households or population
	Affordability	SOC2	Share of household income spent on fuel and electricity	– Household income spent on fuel and electricity – Household income (total and poorest 20% of population)
	Disparities	SOC3	Household energy use for each income group and corresponding fuel mix	– Energy use per household for each income group (quintiles) – Household income for each income group (quintiles) – Corresponding fuel mix for each income group (quintiles)
Health	Safety	SOC4	Accident fatalities per energy produced by fuel chain	– Annual energy produced

Source IAEA (2005)

Safety, which covers accident fatalities caused by the extraction, conversion, transmission/distribution and use of energy.

The Social indicators for energy sustainability, developed by the IAEA (2005), are represented in Table 3.17, with their respective sub-themes and components.

References

Afgan NH, Carvalho MG (2008) Sustainability assessment of a hybrid energy system. Energy Policy 36(8):2903–2910

Afgan NH, Carvalho MG, Hovanov NV (1999) Energy system assessment with sustainability indicators. Energy Policy

Brundtland Commission (1987) Our common future: report of the World Commission on Environment and Development. UN Documents Gathering a Body of Global Agreements, Oslo

de Carvalho JRM, Curi WWF, de Araújo Carvalho EKM, Curi RC (2011) Proposta e validação de indicadores hidroambientais para bacias hidrográficas: estudo de caso na sub-bacia do alto curso do Rio Paraíba, PB. Revista Sociedade e Natureza, Uberlândia

Goodland R (1995) The concept of environmental sustainability. Annu Rev Ecol Syst 26(1):1–24
Goldemberg J, Lucon O (2007) Energia e meio ambiente no Brasil. Estudos Avançados 21(59):7–20
Harris JM (2003) Sustainability and sustainable development. Available at http://isecoeco.org/pdf/susdev.pdf. Accessed 14 Oct 2017
International Atomic Energy Agency (IAEA) (2005) Energy indicators for sustainable development : guidelines and methodologies. International Atomic Energy Agency, Vienna
Santos TCC, Câmara JBD (Orgs.) (2002) GEO Brasil, 2002: Perspectivas do Meio Ambiente no Brasil. Brasília: Edições IBAMA, 2002
United Nations (2007) Indicators of sustainable development: guidelines and methodologies, 3rd edn. New York
van Bellen HM (2006) Indicadores de Sustentabilidade: Uma Análise Comparativa, 2nd edn. FGV, Rio de Janeiro

Chapter 4
Sustainability Indicators for the Performance Evaluation of Renewable Energy Technologies

Abstract This chapter is focused in the selection of sustainability indicators for renewable energy evaluation. It compares the most used indicators found in the literature and defines a classification. Finally, a smaller collection of indicators is chosen to be used in the case study presented in Chap. 6. A detailed definition is described for each indicator.

Considering the sustainable development and energy sustainability indicators presented in Chap. 3, as well as the importance of these indicators and the procedure to select them, this chapter brings some of the most relevant sustainability indicators for the application on the performance evaluation of renewable energy technologies, unfold in Chap. 5. From now on, the focus will be on the most relevant technologies with regard to evolution of this subject in Brazil for larger investments for the use of communities, i.e.: hydroelectric, wind and photovoltaic sources. In any case, the use of energy storage systems (ESS) using batteries are still not considered since this technology is still expensive and its market is under development globally and marginal in Brazil.

As these indicators are more specific for analyzing renewable energy technologies with regard to economics and their impact in the environment and society, they were selected here based, not only on the ones presented on Chap. 3 but also on the study of several other scientific materials that developed analysis on the sustainability of renewable sources, especially considering their life cycle and potential impacts. After their selection, this Chapter also gives more details about how to determine or calculate each named indicator, being it quantitative or qualitative.

4.1 Indicators Selection

In order to analyze the sustainability of renewable energy technologies, indicators that are more specific to their development and potential impacts must be selected. Although the sustainable development (CSD indicators) and energy sustainability

J. F. G. de Oliveira and T. C. G. Trindade, *Sustainability Performance Evaluation of Renewable Energy Sources: The Case of Brazil*,
https://doi.org/10.1007/978-3-319-77607-1_4

indicators (ISED) presented in Chap. 3 are of great importance for the evaluation of a Nation's sustainability, based on the Agenda 21 principles, some of them are quite broad, making it difficult to apply directly for the evaluation of specific renewable energy endeavors. With this in mind, in the context of this book the authors expanded the universe of indicators to find a more adequate set. To do this, 24 scientific materials on the sustainability of alternative energy resources were analysed and the most common indicators used by them to measure the impacts were selected. Among these materials are scientific articles, international reports and books that studied in detail how sustainable are the most popular renewable energy sources exploited today, based on life cycle analysis and environmental impact assessments.

Table 4.1 presents the most common indicators used in these studies, as well as their frequency of occurrence. The indicators deemed as the most relevant were those which appeared in more than 20% of the analyzed studies. These indicators are the ones selected for the sustainability performance evaluation methodology for this book, being described in more details in the next sections. The selected indicators are then arranged in the different dimensions of sustainability considered by the methodology (economic, technical, environmental and social) and classified as quantitative or qualitative, as shown in the diagram of Fig. 4.1.

It is evident from Table 4.1 that the emission of greenhouse gases is the most common indicator applied in the sustainability evaluation of energy sources, which shows how climate change has gained expressive international attention. It is clear that the quest to find different energy sources that do not emit greenhouse gases in the same intensity as traditional fuels and are less harmful to the environment is one of the primary reasons of the increasing amount of research and resources that are being directed to the development of alternative energy technologies.

As already stated, the indicators that were present in more than 20% of the analyzed studies were selected for the development of a sustainability evaluation of renewable energy technologies. To facilitate their comprehension and management, these indicators were classified in four different dimensions of sustainability (economic, technical, environmental, and social) according to their characteristics and what they measure. This classification is represented in the diagram of Fig. 4.1.

In order to gather basic information about the indicators presented in Fig. 4.1, the studied literature was revised and the associated values calculated by these studies for each indicator are shown in Table 4.2. It is important to notice that some values or range of values of the same indicator may vary greatly. This can be explained by the fact that different research may have considered different case studies and the particular characteristics of each of these studies may have interfered in the final results. However, the analysis of Table 4.2 is quite relevant, since it can give us an overview of the impacts associated to different renewable energy technologies, shedding also some light on the range of values that each calculated indicator should result in the application of the method.

It is necessary to point out that there is a much more extensive range of indicators that could be considered in a more detailed study on the sustainability of energy generation technologies. However, the study of all these indicators goes

Table 4.1 Frequency of occurrence of different indicators in the analysed scientific material

Indicators	Occurrence (%)
Implementation cost	21
Price of energy generated	21
Payback time	21
Energy intensity	17
Generation efficiency	25
Assured energy	25
Resource depletion	8
Resource availability	21
Area utilization	29
Lifetime	38
Eutrophication	17
Particles emission	13
Lifetime	25
CO_2 payback time	8
Greenhouse gases emission	79
SO_2 emissions	38
NO_x emissions	38
Annual water consumption	21
Biodiversity loss	21
Dismantling conservation units	21
Interaction of structure with fauna	21
Toxins	25
Population removal	21
Land-use impact	21
Visual impact	21
Noise	25

Sources Ardente et al. (2008), Braciani (2011), Coltro et al. (2003), Dos Santos et al. (2006), Dulanp (2015), Evans et al. (2009), Fadigas (2011), Góralczyk (2013), IPCC (2012), Lenzen and Munksgaard (2002), Lenzen and Wachsmann (2004), Lofthouse et al. (2015), Martinez et al. (2009), Mendes (2005), Ong et al. (2013), Pehnt (2006), Pueyo and Fearnside (2015), Schleisner (2000), Shayani et al. (2006), Sherwani et al. (2010), Tahara et al. (1997), Turconi et al. (2013), Varun et al. (2009a, b), WEC (2004)

beyond the scope of this book, which sought to consider only those of greater relevance to the developed evaluation method. Thus, the selected indicators of sustainability of renewable energy systems, represented in Table 4.2, are further detailed in Sects. 4.2 to 4.5.

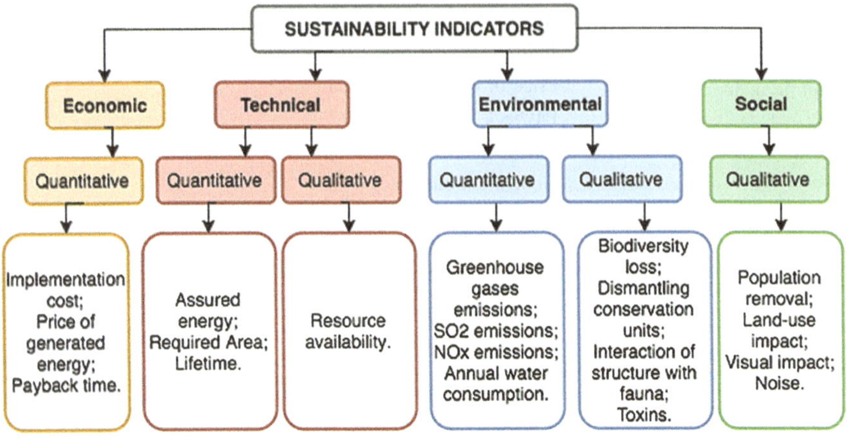

Fig. 4.1 Classification of selected sustainability indicators

4.2 Economic Indicators

Every new energy infrastructure project to be implemented has to be economically viable in the first place. This is the most studied aspect of any investment project. One of the major current concerns regarding the insertion of the new renewable energy generation technologies in the market is whether they can be competitive with traditional and already consolidated energy generation technologies. In several developing countries, notwithstanding their large potential to use renewable energy sources, there are still market barriers to be overcome for the effective commercial implementation of generation technologies. The production scale of these new technologies is one of the great challenges for their effective implementation, and this is due to the small size of the industrial parks and the existence of a restricted market, causing the installation costs to remain high.

The widely used economical parameters for economic decisions in companies are the net present value (NPV) and the internal return rate (IRR). These two indicators are quite popular, however their value is normally not available for public access, for the already finished private investments. Also, since most of the investments evaluated in this book were made by governments, the concern with investment cost, energy price and payback time became quite valuable. In this book, the authors recommend the use of NPV and IRR, but, since the main contribution here is the introduction of the environmental and social aspects, it will only focus its analysis on cost of investment, energy price and payback time.

Taking these aspects into consideration, the following indicators are considered for the study of the economic sustainability aspect of a greenfield energy project:

Table 4.2 Main sustainability indicators of renewable sources found in the studied literature, where HPP is hydro power plant, WPP is wind power plant and PSPP is photovoltaic solar power plant

Sustainability indicators		Unit	HPP	WPP	PSPP
Economic	Implementation costs	USD/kWh	0.05[g]	0.06[g]	0.3[g]
	Price of energy generated	USD/kWh	0.05[e]	0.07[e]	0.24[e]
	Payback time	Years	7–11[m]	0.39–1.4[h]	1.5–15.5[h], 2–5[i]
Technical	Required area	MW/km^2	0.4–15.4[n]	5–10[i]	26.3–45.5[j]
	Lifetime	Years	100[k]	20[a], 20–25[h]	20–30[h]
	Generation efficiency	%	>90[e]	24–54[e]	4–22[e]
Environmental	Greenhouse gas emissions (GGE)	g CO_2 eq/kWh	3.5–6.5[c], 10[d], 41[e], 3.6–11.6[g]	5.2[a], 10.2[d], 25[e], 14.8[f], 7–9[g]	30–80[i], 99[d], 90[e], 98–167[g]
	Emissions from other gases (EOG)	g NOx/kWh	0.004–0.06[b], 0.003–0.006[c]	0.02–0.11[b], 0.015–0.073[c]	0.15–0.40[b], 0.099[c]
		g SO_2/kWh	0.001–0.03[b], 0.009–0.024[c]	0.02–0.09[b], 0.015–0.087[c]	0.12–0.29[b], 0.215[c]
	Water consumption	kg/kWh	36.0[e]	1.0[e]	10.0[e]
	Biodiversity impacts	Qualitative	Modifications in the river course[e], *biodiversity loss*[e], physical barriers of species[i], shredding units conservation	Collision with birds and bats[e], fauna *interaction with wind turbines*[l]	Toxins (production *of plates*)[e]
Social	Social impacts	Qualitative	Population displacement[e], reduction in arable area[e], loss of social and cultural heritage[o]	Noise[e], visual impact[l], land use impact[l], electromagnetic interference[l]	Aesthetic[e], land use impact[i]

Sources [a]Lenzen and Wachsmann (2004), [b]Turconi et al. (2013), [c]WEC (2004), [d]Pehnt (2006), [e]Evans et al. (2009), [f]Ardente et al. (2008), [g]Dulanp (2015), [h]Varun et al. (2009a, b), [i]IPCC (2012), [j]Ong et al. (2013), [k]Coltro et al. (2003), [l]Fadigas (2011), [m]Lofthouse et al. (2015), [n]Dos Santos et al. (2006), [o]Mendes (2005)

4.2.1 Implementation Cost

This indicator is related to the cost of implementing the power plant infrastructure. It depends on the energy that will be explored, but can also vary greatly according to the design of the plant structure (Dunlap 2015). The implementation cost directly interferes with other economic indicators, such as the return on investment time-period and the price of the energy generated, a very important indicator for the economic feasibility of installing a renewable energy generation plant. Some of the typical costs related to the infrastructure required for the studied sources are represented, in USD/kWh installed, in Table 4.2. The implementation cost (IC) of an energy production plant can be calculated, in a simplified form, by dividing the cost of the initial investment by the installed power of the plant (IP), as shown in Eq. 4.1.

$$IC[\mathrm{USD/W}] = \frac{Initial\ Investment[\mathrm{USD}]}{IP[\mathrm{W}]} \tag{4.1}$$

4.2.2 Price of Energy Generated

Each energy technology has different average electricity generation price and a wide range of costs. The average values presented in Table 4.2 were calculated by an extensive literature review, considering the complete life cycle of each technology, bringing together generation costs from construction, installation, operation and maintenance. End of life aspects are also considered, such as cost to dismantling, recycling and/or disposal (Evans et al. 2009). The figures presented are in dollars for each produced kWh.

The price of power generation directly boosts incentives or disincentives for their consumption, production or efficiency improvements (International Atomic Energy Agency (IAEA) 2005). In addition, prices may also affect how much the considered source is accessible or not to the population. In this way, energy prices can be regulated to internalize environmental and social costs, to manage demand and encourage the development of alternative renewable energy options (IAEA 2005). Price profiles for each renewable energy technology are typically influenced by the high demand for capital investment combined with low operating and maintenance costs (Evans et al. 2009). The high cost of photovoltaic solar energy in relation to the other sources, observed in Table 4.1, is directly related to the expensive process of silicon purification for the photovoltaic cell production. On the other hand, wind energy costs vary greatly with the adaptation of the size of the installed generators with the quality of the local wind resource (Evans et al. 2009). For hydropower, the construction of the dam is responsible for almost all the costs internalized in the cost of production, since a hydroelectric plant presents low costs of operation and maintenance, considering its long life.

4.2.3 Payback Time

The concept of payback time, also known as payback, relates to the time an investment will take to be paid from the gains of the project. The return times, in years, considered in this book for each of the studied generation technologies were obtained through the analysis of different sources (IPCC 2012; Lofthouse et al. 2015; Varun et al. 2009a, b), considering that each one presented a different time range due to the different factors involved in each analysis. This indicator is preferred when the investment is done by a government or a public institution or company. In these cases, there is not a strong concern with profit and the cash flow is part of the investment decision. As mentioned before, for private investments, other indicators such as the net present value and the internal return rate are also used.

Usually, the initial investment cost of a renewable energy generation venture is higher than that of a non-renewable energy source. The high initial investment related to these projects makes it necessary to determine the long-term financial viability of the construction and production of energy, in order to guarantee the choice of the most adequate source to be explored in a given region.

This viability is frequently evaluated by the payback time indicator. The simplicity of its calculation and the easy assimilation of its results, makes this indicator one of the most used in energy diagnostics (IPCC 2012). However, the calculation of this indicator, represented in Eq. 4.2, presents some disadvantages, since it does not consider the cash flows that occur after the payback is completed. Moreover, the equation also does not include the effects of the fluctuation of money value over time.

$$PB(years) = \frac{IC}{(AP * V - O\&M)} \tag{4.2}$$

In Eq. 4.2, IC represents the total cost to install the power generation plant (USD), AP represents the annual energy production (MWh/year); V represents the market price of energy (R$/MWh); and O&M represents the annual costs of operation and maintenance, including replacements and renewals over the lifetime of the system (USD/year).

4.3 Technical Indicators

The technical indicators can be defined as those which represent physical, temporal and performance aspects of the evaluated power plant technologies. Examples include: assured energy, availability of the resource exploited, utilization of the area, and lifetime of the project. It is important to emphasize that the technical indicators of a project are closely associated with its economic and environmental aspects.

If a determined area required for a project is used efficiently, more energy will be generated in a smaller space. Consequently, not only the environmental modifications and impacts related to the implementation of this project will be smaller, but the implementation costs will also drop, for less area has to be acquired. In addition, if the resource exploited is abundant in the region, the area to be occupied to explore such resource may also be reduced.

4.3.1 Required Area

this indicator is related to the area required for the implementation, operation and dismantling of a power plant, throughout its lifetime. The area required for each renewable energy source studied was determined through the analysis of different studies (IPCC 2012; Ong et al. 2013; Dos Santos et al. 2006). This indicator is given in Table 4.2 in terms of how many MW of electrical power can be produced through a given source in the physical space of one square kilometer. The calculation of the Required Area (*RA*) can be simplified by the Eq. 4.3, where *IP (*MW*)* represents the installed power of a plant and A (km^2) the total area required by the project.

$$RA = \frac{IP}{\mathrm{A}} \tag{4.3}$$

4.3.2 Lifetime

This indicator measures the duration scope of an energy plant, after its implementation until its dismantling. The lifetime tends to vary very much depending on how robust the technology used for electricity generation is, as seen in Table 4.2, through the data analysis from different sources (Coltro et al. 2003; Lenzen and Wachsmann 2004; Varun et al. 2009a, b). This indicator dictates, for the planning scope, how long the plant will continue to run and produce electricity at a satisfactory rate.

4.3.3 Generation Efficiency

The ability to efficiently transform renewable energy into electricity depends heavily on the technology used in the conversion system. This indicator measures the percentage of electricity that is converted from the utilization of a given energy source, that is, how much energy is actually transformed into electricity in relation to the theoretical calculated power for a particular energy enterprise. The generation efficiency varies widely according to the adopted technology and can reach values higher than 90% for hydroelectric power, about 40% for wind power, and approximately 15% for solar energy (Evans et al. 2009).

4.4 Environmental Indicators

Generally, renewable energies have the main advantage of being less harmful to the environment than traditional non-renewable fossil fuels. However, they should not be assumed to be completely harmless to the environment, or free from negative environmental consequences. Whether during the process of producing the components of the plants, during the implementation, or during the operation, these projects are responsible for environmental impacts, albeit on a much smaller scale when compared to non-renewable sources. It is therefore crucial to quantify or qualify these environmental damages through indicators in order to identify improvement opportunities. Therefore, the life cycle of these ventures can be constantly improved, making them increasingly sustainable. With that in mind, the following indicators are suggested for the environmental sustainability analysis of an energy enterprise:

4.4.1 Greenhouse Gases Emissions

The average values presented in Table 4.2 are related to the calculated greenhouse gas emissions throughout the life cycle of each analyzed renewable energy source, including CO_2 emissions during the production phase of the plant components, and throughout its operation. The emissions, which vary considerably according to the source type studied, are presented in the unit of equivalent CO_2 grams per kWh of produced electricity. This indicator is one of the most recurrent in the literature for evaluating the impacts related to energy generation technologies (Ardente et al. 2008; Dunlap 2015; Evans et al. 2009; IPCC 2012; Lenzen and Wachsmann 2004; Pehnt 2006; WEC 2004), both for both renewable and non-renewable energy sources, because it is directly related to one of the most widespread factors that drives the development of alternative energies today: the concern with global warming and climate changes.

The worst performance of solar energy sources in the quantitative environmental indicators, presented in the Table 4.2, can be explained by analyzing the assessment scope of the studies from which the mean data values were extracted. These studies considered the whole life cycle of a photovoltaic power generation plant, that take into account the raw material exploitation stage of the solar panels. As explained by Dunlap (2015), some of the materials used in the production of photovoltaic modules require substantial energy to produce them. This case is usually related to obtaining relatively rare materials, such as gallium, indium and selenium, common components of photovoltaic units that require extracting large amounts of raw material to obtain a small amount of the element needed (Dunlap 2015). In addition, the manufacturing of solar panels currently consumes substantial energy, which contributes to the increasing gas emission rates and water consumption, when assessing the life cycle of PV solar energy.

4.4.2 *Other Gases Emissions*

Similarly to the greenhouse gases emissions inherent to the life cycle of the production of electricity through any renewable resource, there are also other gas emissions that can be harmful to the environment. This indicator is responsible for measuring the emissions of nitrogen oxides (NOx) and sulfur dioxides (SO_2) related to the production of one kilowatt-hour of electricity from each of the three sources studied (Turconi et al. 2013; WEC 2004).

Both gases, though not related to the greenhouse effect, are known to be harmful both to the environment and to human health. The reaction of NO_x gases in contact with oxygen results in the formation of photochemical smog, a combination of several gaseous and particulate pollutants, responsible for causing eye irritation, reduced visibility, and respiratory diseases (Hinrichs et al. 2010). In addition, these gases are essential for the formation of fine particles and ozone in the soil, both associated with adverse effects on human, animal and plant health. On the other hand, the emission of SO_2 into the atmosphere is the main source of the acid rain phenomenon, responsible for damaging several urban structures, as well as being increasing the disease rate and mortality, when in high concentrations (Hinrichs et al. 2010).

4.4.3 *Water Consumption*

Obtaining data on the quantification of water consumption during the electricity generation is a hard task, especially with regard to renewable energies. This is due to the difficulty in differentiating withdrawn water (water that was collected and later returned to circulation) and consumed water (water from circulation to be consumed inside the plant and not returned to circulation). Therefore, the "water consumption" is regarded as a more adequate indicator to evaluate the sustainability of a system, since it measures the depletion of this resource due to its use for producing energy (Evans et al. 2009). Most of the consumption intrinsic to renewable energies is related to the use of water for the maintenance of the system and production of its components, exhibiting, therefore, low indices when considering the system life cycle, according to Table 4.2.

The reservoir, essential for large hydropower plants, retains huge volumes of water, causing large losses due to the increase in the surface evaporation. The magnitude of this loss varies greatly according to the dam size, volume per square meter and ambient temperature and conditions (Evans et al. 2009). Water is also consumed for the production of photovoltaic modules and wind turbines. However, very small volumes are used during the operation and maintenance of these systems, making the water consumption of their life cycle much smaller. It is possible to infer from Table 4.2 that wind energy presents the lowest water consumption of the considered technologies, followed closely by photovoltaic.

4.4.4 Biodiversity Loss

As discussed earlier, technologies for harnessing renewable energy sources are not exempt from environmental impacts, although less harmful to the environment than conventional sources. Considering that it is very difficult to quantify non-valuables such as biodiversity, this indicator is generally qualitative and reflects the main impacts reported in the literature in terms of the installation and operation of renewable energy plants (Evans et al. 2009; Fadigas 2011; IPCC 2012). The impacts addressed in Table 4.2 are just a few of those already identified by several studies. However, since the technologies used to exploit most renewable energy sources are relatively new, so far there are no comprehensive studies to identify all their long-term impacts on biodiversity.

A qualitative evaluation of different indicators that reflect the impacts to the biodiversity of an enterprise in the place where it will be installed can be accomplished through the criteria established by Noble (2015) and represented in Table 4.3. These criteria are generally used in the Environmental Impact Assessment (EIA) procedures and provide a simplified analysis of the magnitude of a certain impact. However, much more complex analyzes can be carried out for the qualitative evaluation of biodiversity impact indicators and are recommended for more detailed diagnoses on the dimensions of the environmental impacts studied.

As an example, wind power generators may kill a substantial portion of the bat population in a region, that may lead to insect super population and its consequences, such as the transmission of arboviruses to humans.

Table 4.3 Qualitative assessment of the environmental and social indicators

Qualitative value	Type of impact
1	Negligible impact
2	Small impact
3	Moderate impact
4	High impact
5	Severe impact

Source Noble (2015)

4.5 Social Indicators

Just as renewable energies are not free from environmental impacts, their social impacts must also be taken into account. The exploitation of sustainable energy sources results in major gains for society, such as promoting a cleaner and sustainable way of life and diversifying and increasing the security of energy supply. However, despite being responsible for several positive social impacts, this work is focused on evaluating the main indicators of negative impacts on society from these energy sources. Although several social impacts are related to the implementation and operation of power generation systems, the indicators are here selected considering the main impacts commonly reported in the literature for the three energy sources selected to be studied. These indicators are: population displacement, land use, visual and noise impacts.

The construction of high capacity hydroelectric plants or wind farms requires the appropriation of quite large areas. As the potentiality of these energy sources depends largely on where they will be explored, often the ideal area for this deployment is traditionally occupied by a group of people. Relocating these people to a new area involves incalculable losses of non-material goods and interferes directly with the way these people used the land. In addition to the direct social impacts, some externalities that are related to large power plants, often pointed out in studies this subject, are the visual impact and the noise generated by them. In general, the noise pollution happens in wind power plants and it is caused by the turbines and the multiplication mechanisms, which can generate annoyance to the neighborhood and can be even harmful to the health, depending on the intensity of the noise and the time of exposure to it. The visual impact, caused by the abrupt and significant modification of the landscape, can create great opposition from the local population, leading to delays or even to the veto on the construction of such new plants.

Just like the case of the biodiversity impact indicators, it is very difficult to find studies that assign quantitative values to these social impacts, since, although they are also related to material goods, they are directly related to intangible and sentimental aspects, which often cannot be translated into numbers or monetary amounts. Thus, the social indicators appointed in this work were only qualitative, based on the literature (Evans et al. 2009; Fadigas 2011; IPCC 2012; Mendes 2005). These indicators can also be evaluated in a simplified way according to the criteria in Table 4.3.

References

Ardente F, Beccali M, Cellura M, Lo Brano V (2008) Energy performances and life cycle assessment of an Italian wind farm. Renewable and Sustainable Energy Reviews 12 (1):200–217.

Braciani U (2011) Estrutura de Custos para Implantação das Usinas de Geração de Energia Elétrica no Brasil. Florianópolis

Coltro L, Garcia EEC, Queiroz GC (2003) Life cycle inventory of electric energy system in Brazil. Int J Life Cycle Assess Landsberg 8(5):290–296
Dos Santos M, Rosa L, Sikar B, Dos Santos E (2006) Gross greenhouse gas fluxes from hydropower reservoir compared to thermopower plants. Energy Policy 34:481–488
Dunlap RA (2015) Sustainable energy. Dalhousie University, Halifax
Evans A, Strezov V, Evans TJ (2009) Assessment of sustainability indicators for renewable energy technologies. Sidney
Fadigas LAFA (2011) Energia eólica. Manole, Barueri, SP
Góralczyk M (2013) Life-cycle assessment in the renewable energy sector. Krakow
Hinrichs RA, Bach M, dos Reis LB (2010) Energia e meio ambiente. Cengage Learning, São Paulo
Intergovernmental Panel on Climate Change (IPCC) (2012) Renewable energy sources and climate change mitigation: special report of the intergovernmental panel on climate change. Cambridge University Press, Nova York
International Atomic Energy Agency (IAEA) (2005) Energy indicators for sustainable development: guidelines and methodologies. International Atomic Energy Agency, Vienna
Lenzen M, Munksgaard J (2002) Energy and CO_2 life-cycle analysis of wind turbines review and applications. Ren Energy 26(3):339–362
Lenzen M, Wachsmann U (2004) Wind turbines in Brazil and Germany: an example of geographical variability in life-cycle assessment. Rio de Janeiro
Lofthouse J, Simmons RT, Yonk RM (2015) The reliability of renewable energy: hydro. Institute of Political Economy (IPE), Utah
Martinez E, Jimenez E, Blanco J, Sanz F (2009) LCA sensitivity analysis of a multi megawatt wind turbine. Appl Energy 87(7):2293–2303
Mendes NAS (2005) As usinas hidrelétricas e seus impactos: os aspectos socioambientais e econômicos do Reassentamento Rural de Rosana - Euclides da Cunha Paulista. UNESP, Presidente Prudente
Noble BF (2015) Environmental impact assessment: a guide to principles and practice. Oxford University Press, Oxford
Ong S, Campbell C, Denholm P, Margolis R, Heath G (2013) Land-use requirements for solar power plants in the United States. National Renewable Energy Laboratory, Golden
Pehnt M (2006) Dynamic life cycle assessment (LCA) of renewable energy technologies. Heidelberg
Pueyo S, Fearnside PM (2015) Emissões de gases de efeito estufa dos reservatórios de hidrelétricas: Implicações de uma lei de potência. In: Fearnside PM (ed) Hidrelétricas na Amazônia: Impactos Ambientais e Sociais na Tomada de Decisões sobre Grandes Obras, vol 2. Editora do INPA, pp 227–237
Schleisner L (2000) Life cycle assessment of a wind farm and related externalities. Renew Energy 20(3):279–288
Shayani RA, Oliveira M, Camargo IT (2006) Comparação do custo entre energia solar fotovoltaica e fontes convencionais. In: Congresso Brasileiro de Planejamento Energético (V CBPE). Brasília
Sherwani AF, Usmani JA, Varun (2010) Life cycle assessment of solar PV based electricity generation systems: a review. Renew Sustain Energy Rev 14(1):540–544
Tahara K, Kojima T, Inaba A (1997) Evaluation of CO_2 payback time of power plants by LCA. Energy Convers 38:615–620
Turconi R, Boldrind A, Astrup T (2013) Life cycle assessment (LCA) of electricity generation technologies: overview, comparability and limitations. Elsevier, Amsterdam
Varun, Bhat IK, Prakash R (2009a) LCA of renewable energy for electricity generation systems—a review. Elsevier, Amsterdam
Varun, Prakash R, Bhat IK (2009b) Energy, economics and environmental impacts of renewable energy systems. Elsevier, Amsterdam
World Energy Council (2004) Comparison of energy systems using life cycle assessment. London

Chapter 5
Sustainability Performance Evaluation of Renewable Energy Technologies

Abstract A method on how to evaluate sustainability performance is described, starting from the definition of purpose and scope and followed by a data collection, evaluation of cases, selection of indicators, sustainability assessment, and synthesis of results. In the sustainability assessment, the analytic hierarchical process is described in detail for a decision support assessment. The definition of scenarios on priorities, with respect to social, environmental or economical dimensions is then integrated by using non-dimensional indicators.

The sustainability evaluation of renewable energy production technologies in this book is carried out according to the diagram in Fig. 5.1. This model was developed considering the sustainability performance assessment presented by Araújo (2010), and structured in order to obtain, as an end result, an aggregate indicator of sustainability from the different specific selected sustainability indicators. In this chapter, each step of the sustainability assessment methodology will be described in greater detail to better understand its application in the case study about Brazil, presented in Chap. 6.

5.1 Definition of Project Purpose and Scope

The main characteristic of the sustainability performance analysis developed in this book is the possibility of comparing two or more energy systems regarding their sustainability. Therefore, the first step in applying the method is to define which energy ventures will be compared and what is the spatial scope of the analysis. As posed by Noble (2015) for the Environmental Impact Analysis (EIA) process, it is impossible, and even undesirable, for an evaluation process to consider all the impacts, all components and characteristics of the evaluated ventures. This statement can also be considered valid for the Sustainability Performance Assessment. Therefore, the definition of the initial scope of the analysis is essential to determine which parameters are crucial to the evaluation, establishing the spatial and temporal

J. F. G. de Oliveira and T. C. G. Trindade, *Sustainability Performance Evaluation of Renewable Energy Sources: The Case of Brazil*,
https://doi.org/10.1007/978-3-319-77607-1_5

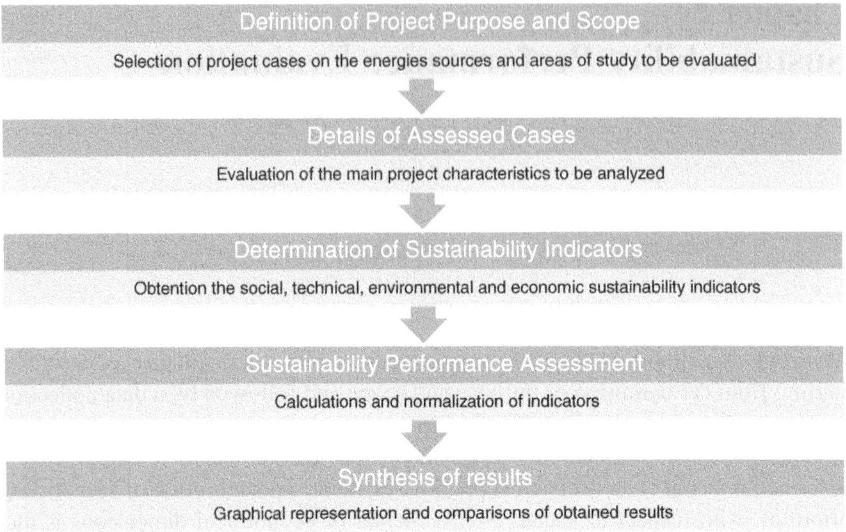

Fig. 5.1 Proposed method for assessing the sustainability performance of renewable energy production technologies

horizon that this analysis will have and focusing on the main concerns and characteristics that can be raised within those horizons (Noble 2015).

For that reason, in order to define the scope of the analysis, the following questions should be taken into account:

- What are the main objectives sought by the application of the method?
- How many energy production technologies do you want to compare?
- Which energy production technologies do you want to compare?
- Which power plants related to those technologies will be studied?
- Where are the studied enterprises located?
- What is the spatial scope of the analysis?
- What is the expected comparison with such spatial scope?
- What will be the analysis' temporal scope?
- Which dimensions of sustainability will be considered in the evaluation?
- Which indicators will be studied within each dimension?

5.2 Data Collection

With the main goals, the energy sources and the spatial and temporal scope of the analysis determined, the next step is to collect data related to real existing projects that will be assessed by the sustainability performance evaluation. This is a way to get the most realistic and reliable values for the analysis. The primary data is essential for the determination of the indicators that will be used in the assessment and can be raised through the study of the characteristics of the analyzed projects in each case study. The necessary data that could not be found directly by the data collection of those enterprises can be estimated based on the general averages of performance presented in the literature, which are normally related to the characteristics of plants that use similar energy sources. When all the essential data is gathered, it is possible then to calculate the indicators to be used in the application of the performance evaluation.

The amount and details in the data that should be collected depend greatly on the scope and precision desired for the performance evaluation that will be developed. However, the following information is essential to support any sustainability performance assessment of renewable energy technologies:

- Physical, climatic and environmental characteristics of the location where the studied projects were installed;
- Availability of the exploited resource;
- Installed power [MW];
- Assured energy (contractual definition of the amount of energy that a power plant generates, corresponding to a fraction of the total generation capacity of the interconnected grid. It is based on this definition that the power plant remuneration is determined, independently of the amount of energy actually generated) [MWh/year];
- Total area, built or not, occupied by the project [km^2];
- Initial investment cost [USD];
- Supply capacity [inhabitants];
- Estimated lifetime of the enterprise [years];
- Annual water consumption [t/year];
- Annual greenhouse gas emissions based on the life cycle of the energy source [kg CO_2eq/year];
- Annual emissions of other gases based on the life cycle of the energy source [kg NOxeq/year and kg SO_2eq/year];
- Price of generated energy power [USD/kW].

5.3 Determination of Sustainability Indicators

From the information gathered by the data collection of the projects that will be evaluated, it is possible to calculate the indicators of each selected sustainability dimension. As suggested in Chap. 4, indicators can be divided into economic, technical, environmental and social. In addition, these indicators can be quantitative, each presenting their specific unit, or qualitative, evaluated by different methods of impact analysis, as suggested in Table 4.2.

Table 5.1 Indicators selected for the performance evaluation methodology

Sustainability indicators		
Type	Indicator	Unit
Economic	Implementation cost	USD/kW
	Price of generated energy	R$/MWh
	Payback time	Years
Technical	Assured energy	MWh assrd./MW
	Resource availability	Qualitative
	Utilization of the area	MW/km^2
	Life-time	Years
Environmental quantitative	Greenhouse gas emissions	kg CO_2eq/MWh
	Emissions of other gases	kg NOx/MWh
		kg SO_2/MWh
	Annual water consumption	t/MWh
Environmental qualitative	Loss of biodiversity	
	Dismantling conservation units	
	Interaction of the structure with the fauna	
	Toxins	
Social	Population displacement	
	Land-use Impact	
	Visual impact	
	Noise	

The main indicators suggested for the sustainability performance analysis methodology developed in this book are represented in Table 5.1. However, as mentioned previously, the number of selected indicators and the accuracy of the data obtained depend very much on the objectives of the analysis and the level of detail needed to reach them. When more indicators are considered, more complete the evaluation tends to be. Obviously, to many indicators may lead to a noise in the analysis as mentioned before.

5.4 Sustainability Performance Assessment

The performance comparison method developed in this book is based on the proposal presented by Araújo (2010) for evaluating the performance of manufacturing processes considering sustainability parameters. This method is based on the calculation of an Aggregate Sustainability Index (ASI), which allows integrating the various studied indicators into a single value, in order to fully capture the performance of the assessed systems and facilitate comparing their level of sustainability (Araújo 2010). According to the Global Report Initiative (GRI) (2006), the main objective of an aggregate sustainability index is to enable the joint analysis of indicators, considering each of the assessed categories in the three sustainability dimensions (GRI 2006).

However, since each one of the selected indicators has different dimensions and units, it is difficult to aggregate them into a common non-dimensional index. Therefore, the indicators must be normalized before the calculation, so that the different categories studied are directly assessed. The calculation of this normalization was carried out using the following formula, adapted from Wenzel, Haushild and Alting (1997):

$$NP(j) = \pm 1 * P(j) * R(j)^{-1} \tag{5.1}$$

Given that *R*(*j*) is the normalization factor, *P*(*j*) is the indicator of the project, and NP(j) is the normalized indicator. The normalization factor *R*(*j*) used was the arithmetic mean of the values analyzed for that indicator. In addition, those indicators that could have a positive impact on the sustainability of the systems evaluated when their values increase, were multiplied by 1. In other words, when normalized, the positive impacts resulted in dimensionless indicators with a positive sign and the negative impacts resulted in dimensionless indicators with a negative sign. Thus, when calculating the Aggregate Sustainability Index (ASI), those with higher values indicate better sustainability performance, while those with lower values indicate a worse performance.

For a better understanding of the application of Eq. 5.1, an example of *NP* (*j*) calculation for three economic indicators of three hypothetical energy sources is given in Table 5.2.

In Table 5.2, the arithmetic means of the given indicators were calculated, and the results represented the normalization factor (*R*(*j*)) in Eq. 5.1. As the Implementation Cost and Payback Time represent indicators that could have a negative economic impact on the sustainability of the systems evaluated when their values increase, they were multiplied by −1. On the other hand, as the market Price of Generated Energy was considered as a positive impact indicator, it was multiplied by 1. The application of Eq. 5.1 to the hypothetical energy sources resulted, then, on the normalized indicators, represented by $NP(S1)$, $NP(S2)$ *and* $NP(S3)$.

With the standardized indicators, the next step for calculating aggregate sustainability indexes is related to determining the weights (W_j), which will be applied in each indicator category, as indicated in Eq. 5.2. These weights correspond to a factor for the impact potential and are determined from the point of view of a decision maker whose task is to find out which energy source should be explored. The stakeholders responsible for promoting renewable energies present different visions and strategic priorities, thus valuing the categories of indicators differently. In other words, investors can associate different weights for each individual indicator depending on their interests. Therefore, in order to facilitate the process of defining weights for the indicators, the application of an analytic hierarchical process (AHP) is recommended.

According to Araújo (2010), this method is based on analysis in pairs, and based in the view of specialists, to derive priority factors. The analysis developed in the

Table 5.2 Example of NP(j) calculation for three hypothetical energy sources

Sustainability Indicators			Source 1 (*S*1)	Source 2 (*S*2)	Source 3 (*S*3)	Arithmetic mean (R(j))	*NP*		
Type	Indicator	Unit					*NP* (*S*1)	*NP* (*S*2)	*NP* (*S*3)
Economic	Implementation Cost	USD/kW	1,000.00	2,000.00	3,000.00	2,000.00	−0.50	−1.00	−1.50
	Price of Generated Energy	R$/MWh	80,00	70,00	75,00	75,00	1.07	0.93	1.00
	Payback time	Years	6	10	14	10	−0.60	−1.00	−1.40

Table 5.3 Calculation of the indices for the priority matrix

A_1	$P_{A(eco)}/P_{A(eco)}$
A_2	$P_{A(eco)}/P_{A(tec)}$
A_3	$P_{A(eco)}/P_{A(env)}$
A_4	$P_{A(eco)}/P_{A(soc)}$
B_1	$P_{B(tec)}/P_{B(eco)}$
B_2	$P_{B(tec)}/P_{B(tec)}$
B_3	$P_{B(tec)}/P_{B(env)}$
B_4	$P_{B(tec)}/P_{B(soc)}$
C_1	$P_{C(env)}/P_{C(eco)}$
C_2	$P_{C(env)}/P_{C(tec)}$
C_3	$P_{C(env)}/P_{C(env)}$
C_4	$P_{C(env)}/P_{C(soc)}$
D_1	$P_{D(soc)}/P_{D(eco)}$
D_2	$P_{D(soc)}/P_{D(tec)}$
D_3	$P_{D(soc)}/P_{D(env)}$
D_4	$P_{D(soc)}/P_{D(soc)}$

priority matrices (Tables 5.5, 5.6 and 5.7) is explained in more detail in Tables 5.3 and 5.4. Recalling that the indices calculated in Table 5.3 are related to the importance of the weight given by the investor to a given sustainability dimension when compared to another dimension. For example, the C_2 index refers to the weight of importance given to the Environmental dimension of a project, when compared to its Technical dimension, according to an investor's point of view. The weight of each dimension was calculated by the arithmetic mean of the different W_j indexes obtained by the analysis, as can be seen in Table 5.4.

The P_x variables, showed in Table 5.3, represent the relative importance given by a certain investor to the sustainability dimension in comparison to the others. These variables have a range that goes from 1 to 10. For instance, if an investor absolutely prioritizes the economic aspects of the project despite its environmental consequences, P_{eco} will be 10 and P_{env} will be 1, resulting in a $A_3 = 10$.

Table 5.4 Priority matrix

	$P_{Economic}$	$P_{Technical}$	$P_{Environmental}$	P_{Social}	Weights (Wj)
$P_{Economic}$	A_1	B_1	C_1	D_1	
$P_{Technical}$	A_2	B_2	C_2	D_2	
$P_{Environmental}$	A_3	B_3	C_3	D_3	
P_{Social}	A_4	B_4	C_4	D_4	
Sum	ΣA	ΣB	ΣC	ΣD	
Weco	$A_1/\Sigma A$	$B_1/\Sigma B$	$C_1/\Sigma C$	$D_1/\Sigma D$	W_{eco} Arithmetic average
Wtec	$A_2/\Sigma A$	$B_2/\Sigma B$	$C_2/\Sigma C$	$D_2/\Sigma D$	W_{tec} Arithmetic mean
Wenv	$A_3/\Sigma A$	$B_3/\Sigma B$	$C_3/\Sigma C$	$D_3/\Sigma D$	W_{env} Arithmetic average
Wsoc	$A_4/\Sigma A$	$B_4/\Sigma B$	$C_4/\Sigma C$	$D_4/\Sigma D$	W_{soc} Arithmetic mean

Fig. 5.2 Scenarios to be developed for the application of the methodology

As each investor type will present his own point of view, resulting in different weights, it is recommended for the development of the Sustainability Performance Evaluation to define some different scenarios. These scenarios aim to guide the process of defining weights for the categories of indicators, taking into account investors with different strategic interests and visions. Here three scenarios are suggested and synthesized in Fig. 5.2. These scenarios are further detailed below, with the respective definition of weights relative to them.

- **First Scenario (CN-mon)**: the investor prioritizes the economic return and the technical and performance characteristics of the power generation system, thus assigning greater value to the performance of economic and technical indicators, to the detriment of environmental and social indicators. The weights assigned to this scenario as a result of the hierarchical analysis are represented in Table 5.5.
- **Second Scenario (CN-eco)**: the investor is very concerned about the performance of environmental and social indicators and is therefore willing to invest in renewable energy generation systems that greatly value these aspects, even if they are not as advantageous in a technical or economic manner. The weights assigned to this scenario are represented in Table 5.6.
- **Third Scenario (CN-sust)**: the investor of this scenario seeks to integrate all three sustainability dimensions, equally valuing all assessed indicator categories. Therefore, this investor has great interest in what characterizes a balanced sustainable production system. The weights assigned to this scenario are represented in Table 5.7.

From the definition of the weights for each dimension of the different scenarios, it is possible to calculate the sub-indexes for each of the categories of indicators

Table 5.5 Priority matrix for CN-mon weights

	$P_{Economic}$	$P_{Technical}$	$P_{Environmental}$	P_{Social}	Weights (W_j)
$P_{Economic}$	1/1	1/1	9/1	9/1	
$P_{Technical}$	1/1	1/1	9/1	9/1	
$P_{Environmental}$	1/9	1/9	1/1	1/1	
P_{Social}	1/9	1/9	1/1	1/1	
Sum	2.22	2.22	20.00	20.00	
W_{eco}	0.45	0.45	0.45	0.45	0.45
W_{tec}	0.45	0.45	0.45	0.45	0.45
W_{amb}	0.05	0.05	0.05	0.05	0.05
W_{soc}	0.05	0.05	0.05	0.05	0.05

Table 5.6 Priority matrix for CN-eco weights

	$P_{Economic}$	$P_{Technical}$	$P_{Environmental}$	P_{Social}	Weights (W_j)
$P_{Economic}$	1/1	1/2	1/5	1/5	
$P_{Technical}$	2/1	1/1	1/5	1/5	
$P_{Environmental}$	5/1	5/1	1/1	1/1	
P_{Social}	5/1	5/1	1/1	1/1	
Sum	13.00	11.50	2.40	2.40	
W_{eco}	0.08	0.04	0.08	0.08	0.07
W_{tec}	0.15	0.09	0.08	0.08	0.10
W_{amb}	0.38	0.43	0.42	0.42	0.41
W_{soc}	0.38	0.43	0.42	0.42	0.41

Table 5.7 Priority matrix for the CN-sust weights

	$P_{Economic}$	$P_{Technical}$	$P_{Environmental}$	P_{Social}	Weights (W_j)
$P_{Economic}$	1/1	1/1	1/1	1/1	
$P_{Technical}$	1/1	1/1	1/1	1/1	
$P_{Environmental}$	1/1	1/1	1/1	1/1	
P_{Social}	1/1	1/1	1/1	1/1	
Sum	4.00	4.00	4.00	4.00	
W_{eco}	0.25	0.25	0.25	0.25	0.25
W_{tec}	0.25	0.25	0.25	0.25	0.25
W_{amb}	0.25	0.25	0.25	0.25	0.25
W_{soc}	0.25	0.25	0.25	0.25	0.25

studied: economic (I_{eco}), technical (I_{tec}), social (I_{soc}) e environmental (I_{env}). The calculation of these sub-indices is done using the following formula (Araújo 2010):

$$I_s = \sum_j W_j * NP(j) \tag{5.2}$$

where, W_j is the weight factor for the impact potential or input indicator j, $NP(j)$ is the normalized indicator and I_s is the impact sub-index. According to Araújo (2010), an obstacle that arises during the integration of indicators is the difficulty in defining adequate weight factors. These factors should be based both on the conditions of the location where the business operates, and on the type of sustainability policy of those responsible for the investment. Thus, it is important that the weighting factors reflect the strategic objectives of the investors, who will attribute different weights to the importance of the economic, technical, environmental and social impacts of the business venture in which they may invest.

For a better comprehension of the application of Eq. 5.2, in Table 5.8 shows an example of the I_{eco} calculation (economic sub-index) for the same three hypothetical energy sources presented in the example of Table 5.2.

In Table 5.8, the $NP(j)$s, previously calculated in Table 5.2, are multiplied by a hypothetical weight assigned to the economic dimension (Wj). The results for each considered indicator were then summed up, determining the economic impact sub-index for the three hypothetical sources. Considering only the I_{eco}, the first source (S1) was the one that presented the higher value. Thus, S1 is the source with the better economic sustainability performance of the three hypothetical sources analyzed.

Finally, to conclude the application of the method, the Aggregate Sustainability Index, ASI, is calculated by the sum of all impact sub-indices (Eq. 5.3), which will result in a single and dimensionless measure that will aggregate the indicators studied and allow assessing the renewable energy systems for power generation in all sustainability axes.

$$ASI = I_{eco} + I_{tec} + I_{amb} + I_{soc} \tag{5.3}$$

With the Aggregate Sustainability Index calculated for each of the renewable energy generation technologies evaluated, it is possible to determine which source has the best performance in each scenario studied. These results can be presented in the form of graphs or tables that compile all the information obtained by the

Table 5.8 Example of the I_{eco} calculation for three hypothetical energy sources

$NP(j)$			Weights (W_j)	$W_j * NP(j)$		
*S*1	*S*2	*S*3	CN-MON	*S*1	*S*2	*S*3
−0.50	−1.00	−1.50	0.50	−0.25	−0.50	−0.75
1.07	0.93	1.00	0.50	0.53	0.47	0.50
−0.60	−1.00	−1.40	0.50	−0.30	−0.50	−0.70
			I_{eco}	−0.02	−0.53	−0.95

developed analysis, aiming to facilitate the comprehension of the data and its application in the decision-making process of which renewable energy technology is the most appropriate for different situations.

5.5 Limitations of the Method

The adoption and application of any research method will inevitably have limitations. Therefore, to better understand the scope and magnitude of the results achieved through the application of a research method, the limitations involved in the evaluation process should be emphasized. For the sustainability performance analysis developed in this work, there are some restrictions, especially those referred to the process of choosing and determining indicators, outlining the scenarios, the spatial coverage of the studied cases, and the number of energy sources evaluated.

Considering that the selection of indicators that constitutes the evaluation method is a complex and difficult process, different results may be achieved by using other indicators. In addition, the results of the method are also conditioned by the prepared scenarios, since different sustainability performances can be reached if, for each scenario, other weights are assigned.

Considering that choosing the indicators for the evaluation method is a complex and difficult process, different results can be achieved by using indicators that differ from those applied by the method developed in this book. In addition, the evaluation results are also conditioned by the scenarios elaborated, since different sustainability performances can be achieved if other weights are assigned for each scenario. The number of sources compared and the spatial scope of the analysis may also limit the outcome of the assessment, especially if this evaluation is used for a decision-making process as to which source is to be implemented in a given location. In this way, it is emphasized once again the importance of the definition of clear objectives, so that the scope of the evaluation is clearly determined in order to reach them.

Therefore, although the results obtained through the method developed in this book can be of great value for the analysis of the energy issue, basing future decisions and planning in this sector, it is important to highlight that the development of further research that apply this method must be continuous, so that the results may be able to increasingly cover other potential renewable sources of energy and expand the regions to be analyzed.

References

Araújo JB (2010) Desenvolvimento de método de avaliação de desempenho de processo de manufatura considerando parâmetros de sustentabilidade (Tese de doutorado). Escola de Engenharia de São Carlos, Universidade de São Paulo, São Carlos

Global Reporting Initiative (2006) Sustainability reporting guidelines. Available at www.aeca.es/comisiones/rsc/documentos_fundamentales_rsc/gri. Accessed 29 Feb 2007

Noble BF (2015) Environmental impact assessment: a guide to principles and practice. Oxford University Press, Oxford

Wenzel H, Hauschild M, Alting L (1997) Environmental assessment of products, vol 1, 2nd edn. Kluwer Academic Publishers, Boston

Chapter 6
Case Study: Evaluation of Renewable Energy Production Technologies in Two Regions of Brazil

Abstract In this chapter, the method of assessing the sustainability performance of renewable energy production technologies, proposed in Chap. 5, is applied to a case study in the Brazilian context, aiming at demonstrating a practical application and also identifying critical aspects for the analysis. Six renewable energy sources for electricity generation in two regions of Brazil (six projects) are selected, deeply described and their indicators are evaluated for 3 different scenarios. The Appendices A, B and C shows detailed information on the calculation of indicators, AIS and a summary of results on radar charts.

In this chapter, the method of assessing the sustainability performance of renewable energy production technologies, proposed in Chap. 5, will be applied to a case study in the Brazilian context, aiming to demonstrate a practical application. This application will follow the steps illustrated in Fig. 5.1, followed by an analysis and discussion of the results and the conclusions derived from the case study. Considering that the core of this chapter is the evaluation of sustainability performance of the main Brazilian renewable energy sources, the research's scope was delimited by the study of Brazilian power generation projects already implemented or in the design phase, as will be further detailed below.

6.1 Definition of the Objectives and Project Scope

This work aims to assess the sustainability of the main renewable energy generation technologies in Brazil. In order to determine the sources analyzed, a study of the current Brazilian energy matrix and energy sources that present great potential for exploration in the country was carried out. In addition, the scope of the work was delimited by the analysis of three different electricity generation projects in two different states of the Federation.

J. F. G. de Oliveira and T. C. G. Trindade, *Sustainability Performance Evaluation of Renewable Energy Sources: The Case of Brazil*,
https://doi.org/10.1007/978-3-319-77607-1_6

According to ANEEL (2016b), the Brazilian electricity matrix is made up of hydroelectric, thermoelectric, thermonuclear, wind and solar power plants, totaling in 2016 an installed capacity of about 148,946 MW. In 2015, a total of 541,408 GWh of electricity was produced, with hydroelectric power accounting for about 76% of this generation. The share of other renewable sources is far less significant, with wind energy accounting for 3.7, 1.2% for biomass, and less than 0.1% for solar photovoltaic (ANEEL 2016b).

However, the growing concern with the environment and with regard to providing drives for a diversified national energy matrix, among other factors, there is a multiplication of cleaner and more reliable energy generation projects. According to ANEEL (2016b), the current electricity expansion scenario in Brazil, consisting of generation projects that are currently being implemented, amounts to 757 plants, totaling 37,139 MW of capacity to be installed. Among these, the source with the largest expansion estimate (42% of the total) is still hydro, with a total of 21 Hydroelectric Plants (HPP) with a little more than 13,400 MW and also 163 Small Hydroelectric Plants (SHP) with additional 2200 MW (ANEEL 2016b).

The photovoltaic plants, with almost 3000 MW, represent 8% of the capacity to be installed. There are 111 projects planned between 2017 and 2019. With regard to wind power plants, by October 2016, more than 2000 MW of installed capacity started their commercial operations. Some 350 projects are in the pipeline, with approximately 350 MW also in operation in 2016 and more than 8000 MW between 2017 and 2021. With respect to biomass-fueled thermoelectric plants, there is a projected increase in installed capacity of about 1927 MW by 2023 (ANEEL 2016b).

Through these data and what was presented in Chaps. 1–3, it can be perceived that hydroelectric energy is the predominant electricity source in the country, and its analysis is crucial to assess the sustainability performance of renewable energy production technologies. The other sources chosen for this assessment were wind and solar, due to the promising expansion horizons they display in the national energy development plans. Furthermore, wind energy is the second most used renewable source for electricity production in the country, and solar energy has become increasingly viable technologically and economically, particularly due to the significant potential of solar incidence on the Brazilian territory and the reducing system prices. Thus, the assessment to be presented was developed for comparing hydroelectric, wind and solar projects.

As already seen in Chaps. 1 and 2, the availability of renewable energy sources depends on the location and amount of energy available, as well as the capability to exploit this resource (Dunlap 2015). Therefore, in order to compare the three chosen sources, achieving results that were more tangible and effective, the scope of the assessment is limited to only two different Brazilian states: Ceará (CE) and Santa Catarina (SC). These states were chosen because they present the three types of electricity plants (solar, wind and hydroelectric), notwithstanding their different availability of energy resources. Therefore, a comparison could be done between projects both within the State as well as interstate projects. The characteristics of the plants analyzed in both States are detailed below.

6.2 Details of the Assessed Cases

As defined in the Objectives and Project Scope, three electricity generation projects were selected from three different renewable sources of energy (hydroelectric, wind and solar) in the two studied Brazilian states (Ceará and Santa Catarina). Table 6.1 contains a general summary of the main characteristics of the plants selected for sustainability performance assessment, and more details on each case are presented in Sects. 6.2.1–6.2.6.

Table 6.1 Main characteristics of the assessed projects

Data	HPP[d, e]	WPP[b, c]	PSPP[f]
Santa Catarina			
Name	Garibaldi	Wind complex Agua Doce	Tubarão
Location	Abdon Batista/Cerro Negro (SC)	Água Doce (SC)	Tubarão (SC)
Details	3 Francis Turbines	86 wind turbines	19,424 panels
Installed power (MW)	191.9	129.0	3.0
Assured energy (MWh/year)	728,000	359,042	12,725
Area occupied (km^2)	26.74	17.20	0.10
Cost of investment (US$)	297,000,000	234,400,000	17,600,000
Supply capability (population)	500,000	380,000	10,000
Ceará			
Data	SHP[a]	WPP[h]	PSPP[i]
Name	Açude Castanhão	Usina de Geração Eólica de Praia Formosa	Grupo Telles
Location	Jaguaribara (CE)	Camocim (CE)	Pindoretama (CE)
Details	–	50 wind generators	9.223 solar panels
Installed power (MW)	5.0	105.0	3.0
Assured energy (MWh/year)	31,500	416,678	12,725
Area occupied (km^2)	325.0	14.00	0.05
Cost of investment (US$)	3,750,000	156,250.000	6,250,000
Supply capability (population)	13,000	310,000	10,000

Sources [a]SEINFRA (2017), [b]Água Doce (2017), [c]IMPSA (2017), [d]Agência Nacional das Águas (2010), [e]Filippim et al. (2013), [f]Tractebel Energia (2014), [g]ANEEL (2016a), [h]Fundação Getúlio Vargas (FGV) (2009), [i]Serviço Nacional da Indústria (SENAI) (2016)

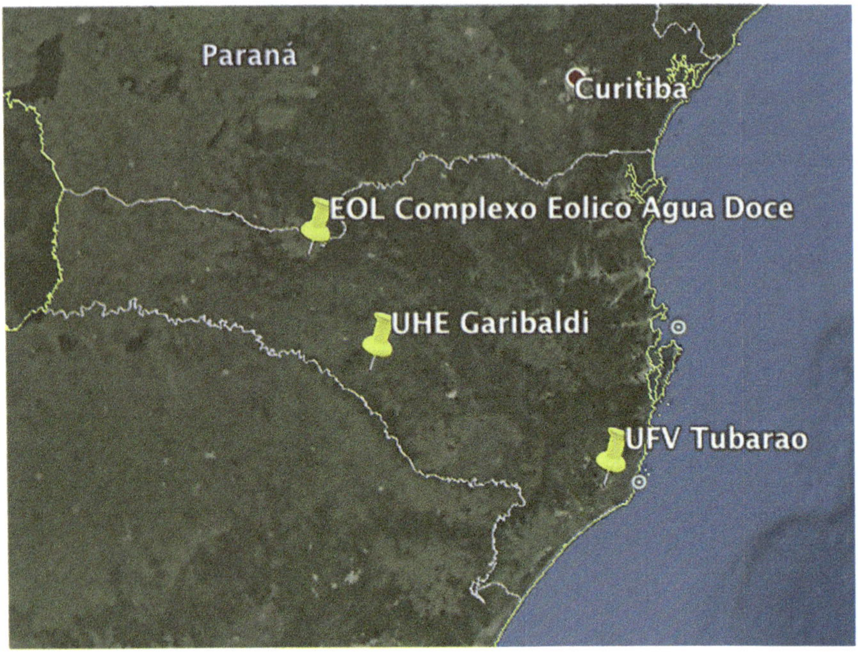

Fig. 6.1 Location of developments in the state of Santa Catarina. *Source* Google Earth (2017)

Figure 6.1 illustrates the spatial location of the plants in the State of Santa Catarina, while the location of the plants in Ceará is represented in Fig. 6.2. Of the studied cases, the SHP of the State of Ceará is the only one among the projects that has not yet been effectively implemented, and is currently only in a project sphere.

It should be noted that most of the data represented in Table 6.1 were extracted from the sources cited there. However, some information was not found in the available literature, such as the total area occupied by the two wind farms, or the energy assured by both solar photovoltaic plants. Thus, these data had to be estimated based on the overall performance/application averages presented by power plants studies using similar energy sources. The total areas occupied by EOLs were estimated according to the average data presented in Table 4.1. However, the energy assured by the PSPPs was estimated considering that they would be able to assure a supply of 50% of their total generation capacity. This assumption was based on the share of the total installed power generally assured by photovoltaic solar plants to ANEEL.

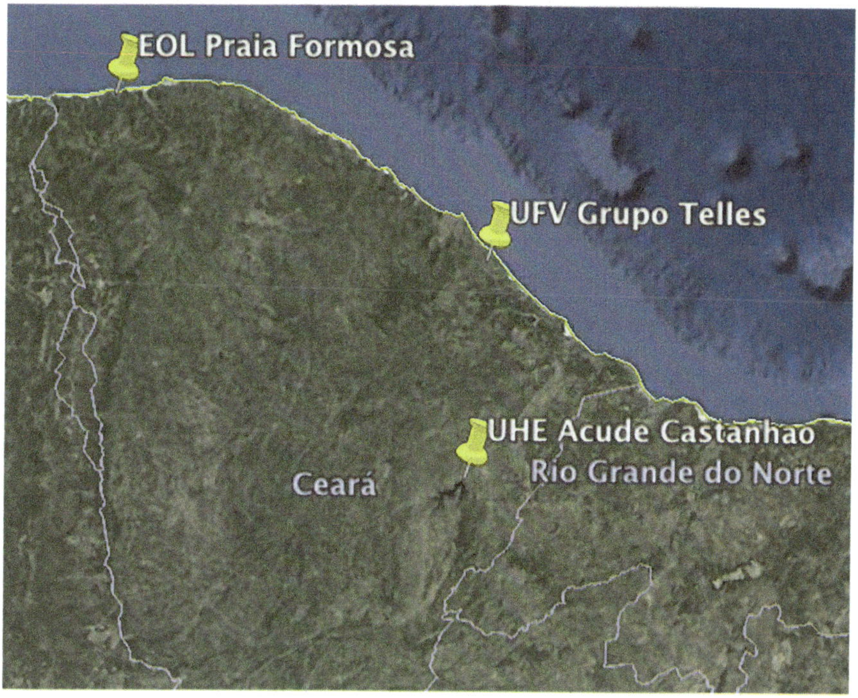

Fig. 6.2 Location of developments in the state of Santa Catarina. *Source* Google Earth (2017)

6.2.1 HPP Garibaldi

The Garibaldi Hydroelectric Power Plant (Fig. 6.3) was built on the Canoas River between the cities of Abdon Batista and Serro Negro—SC in September 2013. With an installed capacity of 191.9 MW, the project has three Francis turbines, and is responsible for generating sufficient energy to supply a population of about 500 thousand inhabitants. The plant, acquired by CTG Brasil in 2015, was granted to Triunfo Energia in 2010, with the right to energy exploration for 35 years, until 2045. Its construction, which lasted about 46 months, had an initial investment of R $ 950 million.

According to the National Water Agency (2010), the river where the HPP was installed begins in the highlands of Santa Catarina, at an altitude of about 1800 m. The Uruguay River is formed along the Pelotas River, its main tributaries are the Caveira and Marombas rivers. The total drainage area of the Canoas River is about 15,000 km^2, and in the dam section of the Garibaldi HPP, the drainage area is of about 13,000 km^2. The hydroelectric power plant arrangement is of the bypass type, with a dam upstream of the reduced flow section (RFS), which is about 14 km

Fig. 6.3 Garibaldi hydroelectric plant—SC. *Source* Prefeitura de Abdon Batista (2017)

long. Its reservoir has a flooded area of 2674 ha, and its main powerhouse is at the end of the bypass, taking advantage of a natural 18 m gap in the RFS (Agência Nacional de Águas (ANA) 2010).

6.2.2 WPP Wind Power Complex Água Doce

Located in the north of the State of Santa Catarina, the Água Doce Wind Complex (Fig. 6.4) is close to the city of Palmas—PR, near the border with the State of Paraná. This complex, developed as part of the PROINFA program by the Argentine company Impsa, consists of six wind farms: Cascata (6 MW), Amparo (22.5 MW), Campo Belo (10.5 MW), Aquibatã (30 MW), Salto (30 MW), and Cruz Alta (30 MW) (IMPSA 2017). The complex, which received a total investment of about 1.3 billion R$, has 86 wind turbines, all IV-77 machines (permanent magnet synchronous generators that couple the blades directly to the generator rotor, avoiding the use of shafts, rotating parts, and the gearbox) of 1.5 MW of unit power each, with a total installed capacity of 129 MW. Its annual electricity production capacity equals the electricity demand of a population of almost 400 thousand inhabitants. In total, the Argentine company Impsa has ten wind farms in Santa Catarina, six units in the Água Doce Wind Complex and four in the Bom Jardim da Serra Complex, providing a total of 222 MW of power in the National Interconnected System (IMPSA 2017).

Fig. 6.4 Água Doce wind complex—SC. *Source* Prefeitura de Água Doce (2017)

In a total area of about 200 km^2, the Água Doce complex was installed directly on the Serra Geral Geological Formation, a member of the Paraná Sedimentary Basin and constituted by extensive volcanic spills (Costanzo et al. 2013). According to Amarante et al. (2001), the general atmospheric flow in this region is controlled by the Depression of the Northeast of Argentina and the Atlantic Subtropical Anticyclone. The pressure gradient between the Northeastern Depression of Argentina and the Atlantic Subtropical Anticyclone induces a persistent northeastward flow throughout the region, resulting in annual wind speeds of 5.5–6.5 m/s. However, this flow is influenced significantly by the terrain's relief and roughness, with the most intense winds (between 7 and 8 m/s) occurring in the highest mountainous elevations of the continent, as well as in low roughness plateaus, as the Campos de Palmas (Amarante et al. 2001).

6.2.3 Tubarão PSPP

The Photovoltaic Solar Power Plant of Tubarão (Fig. 6.5), located in the municipality of Tubarão—SC, started operating in August 2014. Constructed by Tractebel Energia and private and public partners, the plant was designed for training and research with the objective of knowing the costs of investment, operation and maintenance of solar power plants. With an investment of approximately 30 million R$, the energy produced by this PSPP, which has an installed capacity of 3 MW*p*, has the capacity to supply about 2.5 thousand homes (Tractebel Energia 2014).

Fig. 6.5 Tubarão photovoltaic solar power plant—SC. *Source* Tractebel Energia (2014)

The Tubarão Solar Power Plant has a total area of 10 ha, of which 4.5 ha are occupied by its 19,424 photovoltaic panels. These solar modules have three different types of composition, with 1 MW*p* produced by Crystalline Silicon (c-Si) plates, 1 MW*p* by Amorphous Microcrystalline Silicon (a-Si/uc-Si) plates, and 1 MW*p* by CIGS (Copper-Indium-Gallium-Selenium) (Tractebel Energia 2014).

Although the municipality of Tubarão is not the best region of Brazil for a solar power plant, with a daily average solar radiation of 14 MJ/m^2 day, it was strategically chosen because of the available human resources. When compared to the best place for photovoltaic energy generation in Germany, Tubarão, even though it has one of the lowest Brazilian solar radiation indices, has 40% more solar incidence than Germany, which currently has one of the largest installed photovoltaic powers.

6.2.4 Açude Castanhão SHP

The state of Ceará currently has a single SHP, located in the municipality of Varjota, called Araras. This SHP came into operation in the 1960s and is currently shut down for operational convenience. However, the Government of the State of Ceará plans to build a new SHP in the Castanhão Reservoir, which will have an installed capacity of 5 MW, higher than the 4 MW power of Araras SHP. According to the Government of Ceará, this SHP will make use of the water discharge from the Castanhão reservoir, crucial for supplying the transposition axis to the Metropolitan Region of Fortaleza and to the Industrial and Port Complex of Pecém CIPP.

The energy produced will be used by the Water Resources Management Company (COGERH), linked to the Secretariat of Water Resources of the State of Ceará (SRH), in order to optimize the operating costs of the water transfer system (Seinfra 2017). The construction of the SHP will have an approximate investment of 12 million R$, for a projected production of 31.5 GWh/year, with a useful

Fig. 6.6 Açude Castanhão—CE. *Source* National Department against Drought (2017)

falling water of 39 m. Açude Castanhão (Fig. 6.6), officially known as Padre Cícero Public Water, was built on the bed of the Jaguaribe river in the state of Ceará. Its works began in 1995 and were completed in December 2002. With a flooded area of about 325 km^2 and a storage capacity of 6700 million cubic meters, the reservoir is an important mechanism to control the droughts and seasonal floods that reach the Jaguaribe valley, a strategic water reserve for the State. The stored waters are directed to be used in irrigated agriculture, fish farming, fishing, nautical leisure, among other uses (Araújo and Vieira 2001).

6.2.5 WPP Praia Formosa

Inaugurated in September 2009, the Praia Formosa Wind Power Plant had a total investment by Siif Énergies of around R$ 500 million. Located in the municipality of Camocim, Ceará, the wind farm has 50 wind turbines with 2.1 MW of power, with an installed capacity of 105 MW.

The wind power plant, represented in Fig. 6.7, is located in the North-Northeast region of Brazil, which comprises the 100-km-wide coastal strip extending between the northern end of the Amapá coast and the Cabo de São Roque, in Rio Grande do Norte. In this region, the trade winds and the terrestrial breezes are the main influences to wind speeds. According to Amarante et al. (2001), the southern part of this region, which covers the coastal states of Maranhão, Piauí, Ceará and Rio Grande do Norte, have winds with average speeds of 6–9 m/s. These speeds are higher when compared to the rest of the region, since trade winds usually grow stronger as they move away from the Equatorial Depression. In addition, the sea breezes in the South are significantly intensified by lower vegetation and soil moisture levels, raising the surface temperature of the soil during the sunshine hours and, consequently, accentuating the contrast of land-sea temperatures and the resulting sea breeze velocity (Amarante et al. 2001).

Fig. 6.7 Central wind power of Praia Formosa, Camocim—CE. *Source* AVOL – Antonio Viana (2017)

6.2.6 Grupo Telles PSPP

Inaugurated in October 2016, the Telles Group Solar Power Plant (Fig. 6.8) is the largest private solar power plant in the country. The PSPP, which has an installed capacity of 3 MW, is located in Pindoretama, in the metropolitan region of Fortaleza—CE. The plant was implemented in order to meet the Group's exclusive energy needs, which was responsible for installing the equipment. While the project is primarily intended to supply the Santelisa Embalagens carton board packing industry, the generation surplus will be used in other firms of the company (SENAI 2016).

The construction investment for the PSPP was about R$ 30 million, and two-thirds of that amount went to purchase the equipment from the supplier SunEdson, United States. The remainder was invested in electrical, mechanical, civil construction and transformers. About 58 km of electrical cables were buried to interconnect the photovoltaic plant to the industry. In total, this PSPP occupies an area of 6 ha and has 9223 polycrystalline photovoltaic panels, which were produced in Missouri, USA.

Fig. 6.8 Photovoltaic solar power plants in Pindoretama—CE. *Source* Diário do Nordeste (2017)

6.3 Determination of Sustainability Indicators

As defined in the Chap. 3, sustainability indicators are selected parameters and considered separately or combined with each other in order to assess or indicate the performance of a given study object in relation to the three main axes of sustainability: social, economic and environmental. The choice of these indicators is a complex process, which requires certain considerations, since there is such a great diversity, each group being more suited to a given context or universe of analysis (Araújo 2010). To assess the performance of sustainable energy generation technologies, four main groups of indicators were identified in order to cover the main aspects in terms of their implementation and operation. These groups, which will be more detailed, are: economic, technical, environmental and social indicators. Appendix A shows the full data used for the calculation of the indicators.

6.3.1 Economic Indicators

In order to compare the current economic viability of the studied sources with the national market, three economic indicators were established: implementation cost (*USD/kW*), price of energy generated (*USD/MWh*) and period of return on investment (years). The cost of implementation was calculated by the ratio of the investment cost and the installed power, as indicated by Eq. 4.1. As the investment values were found in Reais (*R$:* Brazilian currency), for a better comparison with the international values they were converted to US Dollars (*USD*) through a conversion factor of 3.2 R$/USD. Moreover, the values found for the energy price, established by ANEEL in Auction #9/2015 (ANEEL 2016a), were converted to the US dollar using the same conversion factor. The payback period of the initial investment was calculated by Eq. 4.2.

Table 6.2 represents the economic indicators determined for each of the investments, in the two Brazilian states studied.

It should be noted that the implementation cost, and the consequent initial investment return period, for the SHP in Ceará is considerably lower than the other investments. This is because it is the implementation of an energy generation system in an already built water reservoir. That is, there will be no expenses to implement a new dam, only costs to incorporate an electric power generation system, which significantly reduces the amount of capital that must be invested.

In addition, the large difference between the costs of installing photovoltaic plants in Ceará and Santa Catarina, although these are investments with the same installed capacity, can be explained by the number of solar panels required by each project. While the PSPP of Santa Catarina has 19,424 panels, Ceará has only 9223 panels. This large difference can be attributed to the average daily global solar radiation of Ceará, which is much higher than that of Santa Catarina, requiring less area, and consequently less panels, to produce the same amount of energy.

Table 6.2 Economic indicators of the projects studied

States	Sustainability indicators		HPP	WPP	PSPP
	Indicator	Unit			
SC	Implementation cost	USD/kW	1,547.00	1,817.00	5,865.00
	Price of energy generated	R$/MWh	70,94	69,69	77,81
	Period of initial return on investment	Years	5.75	9.37	15.53
CE	Implementation cost	USD/kW	750.0	1,488.00	2,083.00
	Price of energy generated	R$/MWh	70,94	69,69	77,81
	Period of initial return on investment	Years	1.68	5.38	5.52

6.3.2 Technical Indicators

Table 6.3 shows the technical indicators determined for each of the investments, in the two Brazilian states studied.

The assured energy was determined based on Normative Resolution No. 65, of March 2004, of ANEEL, which specifies the energy assured by the SHPs and WPPs, in addition to other materials published by the National Energy Agency. However, as mentioned earlier, the energy values assured by PSPPs were not found in the literature because these are recent developments. Therefore, these values had to be estimated based on the general performance averages by studies conducted for plants that use similar energy sources (ANEEL 2017; IPCC 2012). For the indicators to be compared to each other, the value of the energy assured by each power plant was divided by its installed power.

The resource availability in each region was determined qualitatively, and the parameters used for this evaluation are represented in Table 6.4. The qualitative values adopted ranged from 1, for little available resource, to 3, abundant resource availability. It should be noted that different evaluation parameters were used for the different evaluated resources, where the water resource was evaluated by the average annual rainfall, the wind resource by the mean wind velocity, and the solar resource by the mean daily global radiation.

The "Utilization Area" indicator was obtained by dividing the installed power by the total area occupied by each project. Thus, the higher the indicator, the more energy that can be produced per square kilometer, which means a better yield of occupied space. As already mentioned, the total area occupied by the two wind farms was not found in the available literature. Thus, these data had to be estimated based on the general performance averages presented by studies for power plants using similar energy sources, with the total areas occupied by the WPPs estimated according to the average data presented in Table 4.2.

The Life-time was estimated according to the means presented by similar undertakings, found in the literature. While hydroelectric plants usually have a life-time of about 100 years, SHPs have a shorter period of about 30 years. On the

Table 6.3 Technical indicators of the projects studied

States	Sustainability indicators		HPP	WPP	PSPP
	Indicator	Unit			
SC	Assured energy	MWh assrd/MW	3,793.6	2,783.3	4,241.7
	Resource availability	Qualitative	3	3	1
	Utilization of the area	MW/km^2	7.18	7.5	30.0
	Life-time	Years	100	20	20
CE	Assured energy	MWh assrd/MW	6,300.0	3,968.4	4,241.7
	Resource availability	Qualitative	1	3	3
	Utilization of the area	MW/km^2	0.015	7.5	60
	Life-time	Years	32	20	20

Table 6.4 Qualitative evaluation of the "resource availability" indicator

Resource	Assessed parameter	Qualitative value	Quantitative value	Source
Water	Average annual rainfall	1	<1100 mm/year	Quadros et al. (2017)
		2	1100 a 1500 mm/year	
		3	>1500 mm/year	
Wind	Average wind speed	1	3–5 m/s	Amarante et al. (2001)
		2	5–7 m/s	
		3	7–9 m/s	
Solar	Daily global solar radiation	1	14–16 MJ/m^2 day	Tiba et al. (2000)
		2	16 to 18 MJ/m^2 day	
		3	>18 MJ/m^2 day	

other hand, solar and wind power plants usually have an approximate life-time of 20 years, also seen in Table 4.2.

6.3.3 Environmental Indicators

Table 6.5 shows the environmental indicators determined for each of the investments, in the two Brazilian states studied.

The quantitative indicators proposed for assessing the environmental performance of the sources evaluated were: greenhouse gas emissions, NO_x emissions, SO_2 emissions, and water consumption. All these indicators were evaluated according to the life cycle analysis of each of the energy sources, especially including the production of system components and the operation of the plants. The calculated values were estimated from the averages of the data found in the studied literature, which are summarized in Table 4.2.

Although there are several indicators that can be used to assess the impacts of difficult-to-assess assets, the qualitative indicators chosen to assess the environmental impacts inherent to the life cycle of the studied sources were: loss of biodiversity, dismantling conservation units, interaction of the structure with the fauna, and release of toxins into the environment. The values assigned to these indicators were determined according to the criteria presented in Table 4.2, based on the Environmental Impact Assessment criteria proposed by Noble (2015).

Moreover, each state had a different performance regarding the qualitative environmental indicators of hydroelectric plants. This is due to the fact that the landscape of each region is different, which interferes directly in the area that has to be flooded for the construction of the reservoir. In Santa Catarina the landscape has more mountains and a completely new reservoir had to be built, while the project in Ceará will be constructed in an already existent reservoir, used for water

Table 6.5 Environmental indicators of the projects studied

States	Sustainability indicators			HPP	WPP	PSPP
	Type	Indicator	Unit			
SC	Quantitative environmental	Annual greenhouse gas emissions	kg CO_2 eq/MWh	12.7	11.87	94
		Annual emissions of other gases	kg NO_x/MWh	0.018	0.0545	0.216
			kg SO_2/MWh	0.016	0.053	0.208
		Annual water consumption	t/MWh	36	1	10
	Qualitative environmental	Loss of biodiversity		5	1	1
		Dismantling conservation units		5	2	1
		Interaction of the structure with the fauna		3	2	1
		Toxins		2	1	3
CE	Quantitative environmental	Annual greenhouse gas emissions	kg CO_2 eq/MWh	12.7	11.87	94
		Annual emissions of other gases	kg NO_x/MWh	0.018	0.0545	0.216
			kg SO_2/MWh	0.016	0.053	0.208
		Annual water consumption	t/MWh	36	1	10
	Qualitative environmental	Loss of biodiversity		3	1	1
		Dismantling conservation units		3	2	1
		Interaction of the structure with the fauna		3	2	1
		Toxins		2	1	3

consumption, and has a flatter landscape, resulting in different environmental impacts magnitudes regarding each case.

6.3.4 Social Indicators

Social indicators were determined qualitatively, following the same guidelines presented in Table 4.2, based on the impact assessment criteria proposed by Noble (2015). These indicators were represented in Table 6.6 for each of the projects, in the two Brazilian states studied.

As with environmental indicators, it should be noted that the performance of the social indicators of hydroelectric plants was different for each State. This is also due to the fact that, as mentioned earlier, in Santa Catarina the project assessed was a hydroelectric plant, while the plant to be installed in Ceará is a SHP, which does not impact the population or the land-use as much as a large hydroelectric plant.

Table 6.6 Social indicators of the projects studied

State	Sustainability indicators		HPP	WPP	PSPP
	Indicator	Unit			
SC	Population displacement		5	1	1
	Land-use impact		5	2	1
	Visual impact		1	3	1
	Noise		2	3	1
CE	Population displacement		3	1	1
	Land-use impact		3	2	1
	Visual impact		1	3	1
	Noise		2	3	1

6.3.5 Method Application in the Cases Assessed

With the indicators determined, it is possible then to begin the environmental sustainability performance evaluation. This section presents how the method detailed in Chap. 5 was applied to the indicators calculated in Sect. 6.3, for each of the assessed projects in the two regions studied.

First, the sustainability indicators were normalized by applying Eq. 5.1, bearing in mind that the value of the normalization factor used corresponds to the average value between the indicator of the three projects in the region studied. The normalized values of the sustainability indicators are shown in Tables 6.7 and 6.8, for the States of Santa Catarina (SC) and Ceará (CE), respectively.

Then, with the normalized indicators, it was possible to develop the three different investor scenarios, proposed in Chap. 5, and the results of each scenario are described below. It is important to notice that Appendix B of this book presents the complete calculation table of the Aggregate Sustainability Indexes of each scenario for the two States covered by the analysis developed by the study.

Finally, with sub-indexes, determined by the application of Eq. 5.2 on the developed scenarios, it was possible to calculate the ASI (Aggregated Sustainability Index) for all the studied sources in all scenarios, enabling the determination of which renewable energy source was the one with the best sustainability performance in each scenario.

6.3.6 Assessment of Scenario 1 (CN-MON)

As previously explained, CN-MON aims to represent an investor's viewpoint, who has great interest in the monetary return and technical performance of the venture, not attributing the same value to environmental and social issues. To represent this

Table 6.7 Normalized values of sustainability indicators for Santa Catarina

Sustainability indicators			HPP	WPP	PSPP	*R(j)*	*NP(j)*		
Type	Indicator	Unit					HPP	WPP	PSPP
Economic	Implementation cost	USD/kW	1,547.00	1,817.00	5,865.00	3,076.00	−0.50	−0.59	−1.91
	Price of energy generated	R$/MWh	70,9	69,70	77,81	72,81	0.97	0.96	1.07
	Return on investment	Years	5.75	9.37	15.53	10.22	−0.56	−0.92	−1.52
Technical	Assured energy	MWh assrd/MW	3,794	2,783	4,242	3,606	1.05	0.77	1.18
	Resource availability	Qualitative	3	3	1	2.33	1.29	1.29	0.43
	Utilization of the area	MW/km^2	7.18	7.5	30	14.89	0.48	0.50	2.01
	Life-time	Years	100	20	20	46.67	2.14	0.43	0.43
Quantitative environmental	Greenhouse gas emissions	kg CO_2 eq/MWh	12.7	11.87	94	39.52	−0.32	−0.30	−2.38
	Emissions of other gases	kg NO_x/MWh	0.018	0.054	0.216	0.10	−0.19	−0.57	−2.25
		kg SO_2/MWh	0.016	0.053	0.208	0.09	−0.17	−0.57	−2.25
	Annual water consumption	t/MWh	36	1	10	15.67	−2.30	−0.06	−0.64
Qualitative environmental	Loss of biodiversity		5	1	1	2.33	−2.14	−0.43	−0.43
	Dismantling conservation units		5	2	1	2.67	−1.88	−0.75	−0.38
	Interaction of the structure with the fauna		3	2	1	2.00	−1.50	−1.00	−0.50
	Toxins		2	1	3	2.00	−1.00	−0.50	−1.50
Social	Population displacement		5	1	1	2.33	−2.14	−0.43	−0.43
	Land-use impact		5	2	1	2.67	−1.88	−0.75	−0.38
	Visual impact		1	3	1	1.67	−0.60	−1.80	−0.60
	Noise		2	3	1	2.00	−1.00	−1.50	−0.50

Table 6.8 Normalized values of sustainability indicators for Ceará

Sustainability indicators			HPP	WPP	PSPP	*R*(*j*)	*NP*(*j*)		
Type	Indicator	Unit					HPP	WPP	PSPP
Economic	Implementation cost	USD/kW	750.00	1,488.00	2,083.00	1,440.00	−0.52	−1.03	−1.45
	Price of generated energy	R$/MWh	70,90	69,70	77,80	72,80	0.97	0.96	1.07
	Return on investment	Years	1.68	5.38	5.52	4.19	−0.40	−1.28	−1.32
Technical	Assured energy	MWh assrd./MW	6.300	3.968	4.242	4.836	1.30	0.82	0.88
	Resource availability	Qualitative	1	3	3	2.33	0.43	1.29	1.29
	Utilization of the area	MW/km^2	0.02	7.5	60	22.51	0.001	0.33	2.67
	Life-time	Years	32	20	20	24.00	1.33	0.83	0.83
Quantitative environmental	Greenhouse gas emissions	kg CO_2 eq/MWh	12.7	11.87	94	39.52	−0.32	−0.30	−2.38
	Emissions of other gases	kg NO_x/MWh	0.018	0.054	0.216	0.10	−0.19	−0.57	−2.25
		kg SO_2/MWh	0.016	0.053	0.208	0.09	−0.17	−0.57	−2.25
	Annual water consumption	t/MWh	36	1	10	15.67	−2.30	−0.06	−0.64
Qualitative environmental	Loss of biodiversity		3	1	1	1.67	−1.80	−0.60	−0.60
	Disassembling conservation units		3	2	1	2.00	−1.50	−1.00	−0.50
	Interaction of the structure with Fauna		3	2	1	2.00	−1.50	−1.00	−0.50
	Toxins		2	1	3	2.00	−1.00	−0.50	−1.50
Social	Population displacement		3	1	1	1.67	−1.80	−0.60	−0.60
	Land-use impact		3	2	1	2.00	−1.50	−1.00	−0.50
	Visual impact		1	3	1	1.67	−0.60	−1.80	−0.60
	Noise		2	3	1	2.00	−1.00	−1.50	−0.50

Table 6.9 Sub-indices and ASI of the Santa Catarina projects

Santa Catarina	CN-MON		
	HPP	WPP	PSPP
I_{eco}	−0.04	−0.25	−1.06
I_{tec}	2.23	1.35	1.82
I_{amb}	−0.58	−0.23	−0.54
I_{soc}	−0.17	−0.20	−0.07
$ASI_{(CN\text{-}MON)}$	1.44	0.66	0.15

point of view, we determined the weights to be used in the scenario from the priority matrix represented in Table 5.5.

With the determined weights, the different categories of indicators were combined, in order to determine the environmental, social, technical and economic sub-index. From their sum, the Aggregate Sustainability Index can be determined. The calculated values for the CN-MON of the State of Santa Catarina are shown in Table 6.9, while those of Ceará are shown in Table 6.10.

For CN-MON of Santa Catarina, according to Table 6.9, the highest performance project was hydroelectric power. It should be emphasized again that this final result was achieved from attributing a much greater weight to the categories of economic and technical indicators, since these would have greater importance than the social and environmental indicators, from the point of view of the investor analyzed. The indicators that most contributed to the higher performance of hydroelectric energy were the return period of the investment and the lifetime of the project.

For the CN-MON of Ceará, according to Table 6.10, the one with the highest performance was solar energy, when a greater weight is attributed to the categories of economic and technical indicators. The indicators that most contributed to the higher solar energy performance were resource availability and area utilization. It is important to emphasize again that the studied SHP, in Ceará, will be implemented in an already built reservoir, which significantly reduces its implementation costs. Therefore, this is one of the reasons why the SHP performed better than WPP, when the indicators are assessed according to the weights set for the CN-MO.

Table 6.10 Sub-indices and ASI of Ceará's projects

Ceará	CN-MON		
	HPP	WPP	PSPP
I_{eco}	0.02	−0.61	−0.76
I_{tec}	1.38	1.47	2.55
I_{amb}	−0.44	−0.23	−0.53
I_{soc}	−0.25	−0.25	−0.11
$ASI_{(CN\text{-}MON)}$	0.72	0.39	1.15

6.3.7 Assessment of Scenario 2 (CN-ECO)

The CN-ECO represents the point of view of an investor whose main concerns are the environmental and social aspects of the venture. Thus, in order to obtain the best possible performance in this respect, this investor would even be willing not to achieve the best technical and economic performance. To represent this point of view, we determined the weights to be used in the scenario from the priority matrix represented in Table 5.6.

The determined weights were then combined to different categories of indicators in order to determine the environmental, social, technical and economic sub-index. From their sum, the Aggregate Sustainability Index was finally determined. The calculated values for the CN-ECO of the State of Santa Catarina are shown in Table 6.11, while those in Ceará are shown in Table 6.12.

For the CN-ECO of Santa Catarina, according to Table 6.11, the project with the highest sustainability performance was wind energy. It should be emphasized that this final result was achieved from attributing a much greater weight to the categories of social and environmental indicators, since these would be more important than the technical and economic indicators, according to the analyzed investor's point of view. For the CN-ECO of Ceará, according to Table 6.12, the one with the highest performance was also wind energy. Moreover, both tables indicated that solar energy obtained the second-best socio-environmental performance, while hydroelectric energy had the worst performance of the studied energy sources.

It is noteworthy that solar energy ventures, despite their low social impacts, had their environmental performance hampered by their life cycle. As mentioned earlier, when the processes of raw material mining and production of solar modules are

Table 6.11 Sub-indices and ASI of the Santa Catarina projects

Santa Catarina	CN-ECO		
	HPP	WPP	PSPP
I_{eco}	−0.01	−0.04	−0.17
I_{tec}	0.50	0.30	0.40
I_{amb}	−3.89	−1.72	−4.23
I_{soc}	−2.30	−1.84	−0.78
$IAS_{CN\text{-}ECO}$	−5.71	−3.29	−4.77

Table 6.12 Sub-indices and ASI of the Ceará projects

Ceará	CN-ECO		
	HPP	WPP	PSPP
I_{eco}	0.00	−0.10	−0.12
I_{tec}	0.31	0.33	0.57
I_{amb}	−3.60	−1.89	−4.35
I_{soc}	−2.01	−2.01	−0.90
$IAS_{CN\text{-}ECO}$	−5.30	−3.66	−4.81

taken into account in the analysis of the solar energy life cycle, this results in a considering greater number of atmospheric pollutants emitted to produce 1 MWh of this energy.

6.3.8 *Assessment of Scenario 3 (CN-SUST)*

Considering that sustainable development can be understood as a development based on the balance between the economic, environmental and social spheres, the CN-SUST was prepared based on an investor's point of view to promote the sustainable production of energy, in all its aspects. Therefore, this investor would give equal importance to all categories of assessed indicators. To represent this point of view, we determined the weights to be used in the scenario from the priority matrix represented in Table 5.7.

As in the other scenarios, the weights determined were combined with the different categories of indicators in order to determine the environmental, social, technical and economic sub-index. From their sum, the Aggregate Sustainability Index was finally determined. The calculated values for the CN-SUST of the State of Santa Catarina are shown in Table 6.13, while those in Ceará are shown in Table 6.14.

For CN-SUST, both in Santa Catarina and Ceará, the project with the highest sustainability performance was wind energy. It should be emphasized that this final result was achieved from attributing equal weights to all categories of assessed indicators, considering that the investor advocates the balance of all sustainability

Table 6.13 Sub-indices and ASI of the Santa Catarina projects

Santa Catarina	CN-SUST		
	HPP	WPP	PSPP
I_{eco}	−0.02	−0.14	−0.59
I_{tec}	1.24	0.75	1.01
I_{amb}	−2.37	−1.05	−2.58
I_{soc}	−1.40	−1.12	−0.48
$IAS_{CN\text{-}SUST}$	−2.56	−1.56	−2.63

Table 6.14 Sub-indices and ASI of the Ceará projects

Ceará	CN-SUST		
	HPP	WPP	PSPP
I_{eco}	0.01	−0.34	−0.42
I_{tec}	0.77	0.82	1.42
I_{amb}	−2.19	−1.15	−2.65
I_{soc}	−1.23	−1.23	−0.55
$IAS_{CN\text{-}SUST}$	−2.64	−1.90	−2.21

spheres. It is important to note that while in Santa Catarina hydroelectric power was the one with the second-best sustainability performance, solar energy was the second best in Ceará. This may reflect the different availability of the two resources in the two regions studied, which greatly influences the technical and economic aspects of the analysis.

6.4 Summary of Results

In this item, the last step of the procedure proposed in the diagram of Fig. 5.1 is presented for the evaluation of the sustainability performance of the renewable energies studied. In this stage, it is intended to condense all the information obtained by the analysis developed in the previous items, in order to provide subsidies for decision making in relation to which renewable energy generation project is the most appropriate for different situations. The ranking of sustainability performance for each of the evaluated scenarios is represented in Table 6.15.

In order to better visualize these results, the graphs of Appendix C (Figs. C.1–C.6) were prepared for the three scenarios analyzed. Figures C.1 and C.2, for the two regions studied, illustrate the CN-MONs which took an investor's point of view that prioritizes the economic and technical performance of his venture. In this scenario, while hydroelectric energy indicated the best performance in Santa Catarina, solar energy was the best in Ceará.

For the two regions studied, the CN-ECO is illustrated in Figs. C.3 and C.4, in which an investor's point of view was taken to prioritize the socio-environmental performance of his venture. In this scenario, wind energy was the one with the best performance for both Santa Catarina and Ceará.

Finally, Figs. C.5 and C.6 illustrate the CN-SUST, for the two regions studied, taking an investor's viewpoint that prioritizes the sustainable performance of his venture, with equal emphasis on all sustainability spheres. In this scenario, both Santa Catarina and Ceará had the best wind energy performance.

Table 6.15 Sustainability performance ranking in the assessed scenarios

Scenario	Performance ranking	Santa Catarina	Ceará
CN-MON	1st	HPP	PSPP
	2nd	WPP	HPP
	3rd	PSPP	WPP
CN-ECO	1st	WPP	WPP
	2nd	PSPP	PSPP
	3rd	HPP	HPP
CN-SUST	1st	WPP	WPP
	2nd	HPP	PSPP
	3rd	PSPP	HPP

Therefore, it is important to point out that wind energy presented the best sustainability performance for both CN-ECO and CN-SUST for the two regions studied. The reasons that led to this result will be discussed in more detail in the next section, as well as the consequences and limitations of this analysis.

6.5 Discussion

Table 6.16 summarizes the results achieved by assessing the sustainability performance of the six renewable energy generation projects analyzed in Santa Catarina and Ceará. Through the analysis of this table, it is indisputable that there is a strong emphasis on wind energy, which presented the best performance in four of the six scenarios studied. As already seen, wind energy occupies third place in the share of the Brazilian electrical matrix, which is the second most used renewable source in the country, after hydroelectric power.

The wind power generation projects presented the best sustainability performance for both states, both in Scenario 2 (CN-ECO), which represented the viewpoint of an investor who shows major concerns about environmental aspects and in Scenario 3 (CN-SUST), which illustrated the case of an investor interested in balancing all aspects involved in the concept of sustainability. These results demonstrate that, within the limitations of the analysis method, the wind energy generation system is the one with the best balance between the advantages and disadvantages of its economic, technical, environmental and social aspects. As a consequence, it is not surprising that wind energy ventures are increasingly growing in Brazil. According to the ranking published by the Global Wind Energy Council (GWEC), an international wind energy organization, in 2016 the country saw an expansion of 2014 Megawatts in the generation of this energy. This caused Brazil to reach the 5th position in the world ranking of installed capacity of wind energy that same year.

In the scenarios analyzed, wind energy was not the best economic option in any of the cases (Scenario 1—Santa Catarina and Ceará). According to Fadigas (2011), the low density of the wind resource, coupled with the lower efficiency of the wind turbine, means that it must have larger size for capturing wind energy, consequently making the equipment more expensive. Thus, the great challenge faced by consolidating wind energy in the market is to further reduce the costs of wind turbines. Regarding the specific case of Brazil, the long-term stabilization of the wind power sector is extremely important for reducing these costs. This can be achieved through

Table 6.16 Higher sustainability performance projects for different scenarios

Scenario	Santa Catarina	Ceará
CN-MON	HPP	PSPP
CN-ECO	WPP	WPP
CN-SUST	WPP	WPP

long-term planning with well-defined rules for this sector and the production of amounts of energy that favor scale economies (Fadigas 2011). Regarding the environmental and social impacts of a wind farm project, these are mainly concentrated in the preparation phase of the location and the installation of wind turbines. Moreover, during its operation, a wind farm may also cause negative impacts related to the interaction of fauna with turbines, visual and noise impacts, and the effects of electromagnetic interference.

Regarding hydropower, as observed in the analyzed indicators, this represents great advantages in relation to the economy of the project, in which the water resource is abundant (Scenario 1—Santa Catarina). This is due, among other factors, to the fact that hydroelectric energy exploration technology is already consolidated in the market, it is robust and is used traditionally and historically in the Brazilian context of electricity production. However, this type of energy use, especially large scale utilization, can be responsible for significant environmental and social damages related to the flooding of large areas to form the reservoir. Among the several impacts of this exploitation, besides altering the natural regime of the river in which the dam was installed, there is the great loss of biodiversity, the displacement of traditional populations and the significant losses of important elements of its natural, cultural and social heritage. However, notwithstanding all these impacts, the source with the largest expansion projection in the Brazilian energy scenario remains the water source, which leads to discussing the need to diversify the national electricity generation model and the need for planning hydroelectric projects that are smaller in size or less impacting to the environmental and social environments, in which they are inserted.

As far as solar energy is concerned, it was found, through the performance analysis carried out, that this is a very interesting project from the economic point of view for regions where there is intense solar radiation incidence (Scenario 1—Ceará). However, this type of energy can be very expensive when implemented in a region that does not have not such an intense resource (Scenario 1—Santa Catarina), because generating a given electric power in this location would require a much greater area as well as numerous photovoltaic solar panels, significantly increasing the costs of this project.

Solar photovoltaic plants have minimal impacts during their installation and operation. However, as discussed during the performance analysis, when the life cycle of solar modules is taken into account, it undermines the environmental performance of this form of energy. This is due to the fact that some of the raw materials used to produce these modules require high energy consumption for their extraction. In addition, the current process for the manufacture of solar panels requires high technology and also consumes substantial energy, which contributes to increasing, during its life cycle assessment, the gas emission and water consumption indices. Therefore, to effectively consolidate this renewable source in the energy market and so that it becomes increasingly sustainable, research investment is crucial for the development of more efficient solar technologies, which contain less impactful life cycles, and whose production processes are simpler, less costly and less energetically intense.

It is also important to highlight, once more, the limitations involved in the sustainability assessment performance process developed throughout this work. As already discussed, selecting the indicators to use as the basis for this analysis is a complex and difficult process. Different results could have been achieved if other indicators had been used, or if other weights had been assigned for each scenario. However, the indicators used were chosen based on the existing literature regarding the sustainability and life cycle assessment of renewable energy generation systems. Although some of the assessed data were not available or could not be found, for some of the parameters of the projects studied, these were estimated based on other studies performed for the sources surveyed, or extrapolated from the performance averages presented by similar generation ventures.

Another aspect that should be raised is that the study was conducted only for two states in different Brazilian regions. As discussed, the availability of renewable energy sources strongly depends on the location and amount of energy available, as well as the capacity to exploit this resource (Dunlap 2015). Thus, wind energy may have achieved a good performance analysis, since it was performed in two regions that have significant abundance of wind resources. However, if the same analysis had been carried out in the Amazon region, which concentrates a large part of the hydroelectric potential not yet explored in the country, but which on the other hand has low wind potential, perhaps the results of the three scenarios analyzed would have been completely different and the hydroelectric power could result in that of greater prominence.

Furthermore, it should also be mentioned that this work assessed only three of the various renewable sources that have exploration potential in Brazil and in the world. As an example, one of the energy sources that has not been addressed and has gained increasing prominence in the world energy panorama is biomass. If the analysis developed in this work had been carried out in the Southeast and Center-West regions, which have significant agricultural production, and the sustainability performance of biomass had been taken into consideration, perhaps it would also receive relative importance in the scenarios surveyed.

Therefore, although the results achieved through the sustainability performance evaluation carried out in this study may be significant to analyze the issue of electricity production from renewable sources in the country and to support future decisions and planning in this sector, it should be emphasized that more research has to be developed both in relation to other renewable sources, as well as to other Brazilian regions. Thus, the particular characteristics of each analyzed region can be considered in a more effective and concrete manner, to facilitate identifying its different potentials and limitations. It is also worth mentioning that preparing more complex scenarios, involving more parameters and indicators, would lead to more reliable and realistic results. Consequently, these results could provide a more objective basis for choosing the most advantageous renewable energy sources for a given region, both in terms of technical and economic aspects, as well as considering environmental and social issues.

6.6 Conclusions

Considering the issue of national energy, promoting a clean and diverse electrical matrix is essential to guarantee the country's sustainable development and to carefully meet its current and future energy demands. The availability of renewable energy sources has particular characteristics depending on the area studied, so the plans for the development and exploitation of these sources should be based on sustainability analyses of each of the regional energy potentials, in order to determine which energies are the most interesting from a technical, economic, social or environmental point of view.

To analyze the sustainability performance of three renewable energy sources (solar, wind and hydroelectric) that present great potential in the Brazilian scenario, this work carried out a performance assessment of two Brazilian States with different characteristics: Ceará and Santa Catarina. Therefore, a dimensionless sustainability index was calculated from sustainability indicators in different scenarios, which represented the point of view of investors with various interests. These indicators allowed to determine which energy sources could be considered more interesting according to each context and region analyzed.

Although the assessment method used had several limitations, especially regarding the choice of indicators, access to information and the scope of the analysis, the results obtained allowed observing the remarkable emphasis of wind energy. This assessment method showed the best sustainability performance in four of the six scenarios studied. Therefore, it can be concluded that the wind energy generation system, in the evaluated context, demonstrated a significant balance between the positive and negative aspects of its economic, technical, environmental and social aspects. Consequently, the progressive growth in number and power of wind energy ventures in Brazil can be justified by these systems' greater sustainability, not only with regard to technical and economic performance, but also socio-environmental performance.

With regard to hydropower, it is the predominant electricity source produced in Brazil. However, despite its more robust and consolidated technology, it lags behind in the sustainability assessment, when the socio-environmental aspects of its implementation are taken into consideration. This is due to the fact that, especially when large scale, the installation of a hydroelectric plant can cause significant environmental and social damages due to the flooding of large areas for the reservoir. Regarding photovoltaic solar energy, although this causes minimal impacts during its installation and operation, its sustainability performance is impaired when the life cycle of the solar modules is taken into consideration.

Therefore, through this work, it was possible to conclude that, although considered a clean energy source, renewable sources are not free from negative impacts. In addition, an energy source may present greater or lesser sustainability relative to other sources, depending on the region, indicators, and investors considered in the analysis. Thus, it is clear that the answer to the Brazilian energy question does not lie in investing in a single renewable source, as has been done for many years in relation to hydroelectric energy. The development of a reliable, clean and safe electrical matrix greatly depends on studying the potentials of the different regions in the country, and investing in its various sources. Therefore, these sources can complement each other so as to compose an electrical system capable of guaranteeing the current and future demands, without compromising the quality of life of consumers or harming the environment in which it is inserted.

6.6.1 Outlook

As previously discussed, this case study was limited to the analysis of only three energy sources in two Brazilian Federal Units. Therefore, subsequent relevant topics for future studies are:

- **Biomass sustainability performance analysis**: although this energy source is also very important in the Brazilian scenario, mainly in the South and Southeast, it was not included in this analysis, but it represents a very important subject, and its sustainability can be assessed in relation to the other sources discussed in this work.
- **Analysis of more indicators and scenarios**: the inclusion of more performance sustainability indicators and the development of more investment scenarios can make the research carried out in this work even more representative, with more concrete and expressive results of the Brazilian energy reality.
- **Analysis of other Brazilian states**: although this work has been limited to only two Brazilian Federative Units, the development of this same analysis in other regions can bring different results, composing a significant framework for the groundwork and orientation of future plans, policies and projects of renewable energy in the country.

References

Agência Nacional de Águas (ANA) (2010) Declaração de Reserva de Disponibilidade Hídrica para o aproveitamento hidrelétrico Garibaldi, localizado no rio Canoas. Nota Técnica Conjunta ANA – DRHI-SDS-SC. ANA

Agência Nacional de Energia Elétrica (ANEEL) (2016a) Últimas Notíciais: Leilão n° 9/2015 é homologado parcialmente. Available at http://www2.aneel.gov.br/aplicacoes/noticias/Output_Noticias.cfm?Identidade=9060&id_area=90. Accessed 22 Feb 2017

Agência Nacional de Energia Elétrica (ANEEL) (2016b) Boletim de Informações Gerenciais 2016. Available at http://www.aneel.gov.br. Accessed 27 Feb 2017

Agência Nacional de Energia Elétrica (ANEEL) (2017) Empreendimentos em Operação. Available at http://www.aneel.gov.br/dados/geracao. Viewed 20 Feb 2017

Amarante OAC, Brower M, Zack J (2001) Atlas do Potencial Eólico Brasileiro. MME/ELETROBRÁS/CEPEL

Araújo JB (2010) Desenvolvimento de método de avaliação de desempenho de processo de manufatura considerando parâmetros de sustentabilidade (Tese de doutorado). Escola de Engenharia de São Carlos, Universidade de São Paulo, São Carlos
AVOL – Antonio Viana. Available at http://www.antonioviana.com.br/. Accessed 5 Feb 2017
Costanzo A, Villarreal M, Freire PE, Lima MA, Pane E, Domingues S (2013) Parque Eólico Água Doce Medições de Resistividade do Solo e de Resistências de Aterramento. Brazil Wind Power Conference and Exhibition
Departamento Nacional de Obras Contra as Secas (2017) Castanhão. Available at http://www.dnocs.gov.br/barragens/castanhao/castanhao.html. Accessed 10 Feb 2017
Diário do Nordeste. Available at http://diariodonordeste.verdesmares.com.br/. Accessed 10 Feb 2017
Dunlap RA (2015) Sustainable energy. Dalhousie University, Halifax
Fadigas LAFA (2011) Energia eólica. Manole, Barueri
Filippim ES, Mocelim JP, Fischer A, Bernardy RJ (2013) As perspectivas de atingidos por hidrelétricas: O caso da construção da Usina Garibaldi. Revista Pensamento & Realidade
Fundação Getúlio Vargas (FGV) (2009) Canal de Energia: Siif Énergies investe R$ 500 milhões na Central Eólica Praia Formosa. Centro de Estudos em Sustentabilidade. FGV-EAESP. Available at http://gvces.com.br/siif-energies-investe-r-500-milhoes-na-central-eolica-praia-formosa?locale=pt-br. Accessed 20 Feb 2017
Global Wind Energy Council (GWEC) (2015) global wind market report. GWEC, Bruxelas. Available from www.gwec.net. Accessed 10 Aug 2017
Google Earth. Available at www.google.com.br/maps. Accessed 20 Jan 2017
IMPSA. Parque Eólico Santa Catarina – Brasil. Available at www.impsa.com. Accessed 9 Feb 2017
Intergovernmental Panel on Climate Change (IPCC) (2012) Renewable energy sources and climate change mitigation: special Report of the Intergovernmental Panel on Climate Change. Cambridge University Press, New York
Noble BF (2015) Environmental impact assessment: a guide to principles and practice. Oxford University Press, Oxford
Prefeitura de Abdon Batista (2017) Lago da Usina Hidrelétrica Garibaldi. Available at http://www.abdonbatista.sc.gov.br/. Accessed 10 Feb 2017
Prefeitura de Água Doce (2017) Turismo: Complexo Eólico Água Doce. Available at http://www.aguadoce.sc.gov.br/turismo/item/detalhe/2245. Accessed 10 Feb 2017
Quadros MFL, Machado LHR, Calbete S, Batista NNM, Oliveira GS (2017) Climatologia de precipitação e temperatura. CPTEC/INPE. Available at http://climanalise.cptec.inpe.br/. Acesso em 24 mar. 2017
SEINFRA (2017) Pequena Central Hidrelétrica no Castanhão. Governo do Estado do Ceará. Available at http://www.seinfra.ce.gov.br/index.php/energia/pequena-central-hidreletrica-no-castanhao. Accessed 23 Feb 2017
SENAI – Departamento Regional do Ceará (2016) Notícias: Grupo Telles inaugura usina solar em Pindoretama. SENAI/CE. Available at http://www.senai-ce.org.br/99662/fiecnoticias/grupo-telles-inaugura-usina-solar-em-pindoretama. Accessed 20 Feb 2017
Tiba C et al (2000) Atlas solarimétrico do Brasil: banco de dados terrestres. Ed. Universitária da UFPE, Recife
Tractebel Energia SA (2014) Comunicado ao mercado, Florianópolis

Appendix A
Calculation of Indicators

Santa Catarina			
Collected data	HPP[d,e]	WPP[b,c]	PSPP[f]
Name	Garibaldi	Complexo Eólico Agua Doce	Tubarão
Location	Abdon Batista/Cerro Negro (SC)	Água Doce (SC)	Tubarão (SC)
Details	3 Francis turbines	86 wind turbines	19,424 panels
Installed power (MW)	191.9	129.0	3.0
Assured energy (MWh/year)	728,000.0	359,042.0	14,560.0
Occupied area (km^2)	26.7	200.0	0.1
Cost of investment (R$)	950,000,000.0	750,000,000.0	56,300,000.0
Supply capability (population)	500,000	380,000	10,000

Sustainability indicators		Unit	Garibaldi	Agua Doce	Tubarão
Economic	Implementation cost	USD/kW	1,547.00	1,816.90	5,864.60
	Price of generated energy[g]	R$/MWh	227,00	223,00	249,00
	Payback time	Years	5.7	9.4	15.5
Technical	Occupied area	MW/km^2	7.2	0.6	30.0
	Lifetime	Years	100.0	20.0	20.0
	Generation efficiency	%	Not relevant	Not relevant	Not relevant

(continued)

J. F. G. Oliveira and T. C. G. Trindade, *Sustainability Performance Evaluation of Renewable Energy Sources: The Case of Brazil*,
https://doi.org/10.1007/978-3-319-77607-1

(continued)

Sustainability indicators		Unit	Garibaldi	Agua Doce	Tubarão
Environmental	GHG emissions	kg CO_2eq	9,245,600.0	4,261,828.5	1,368,640.0
	Emissions of other gases	kg NO_x	13,104.0	19,567.8	3,145.0
		kg SO_2	11,648.0	19,029.2	3,028.5
	Annual water consumption	t/year	26,208,000.0	359,042.0	145,600.0

Ceará

Collected data	HPP[a]	WPP[h]	PSPP[i]
Name	Açude Castanhão	Usina de Geração Eólica de Praia Formosa	Grupo Telles
Location	Jaguaribara (CE)	Camocim (CE)	Pindoretama (CE)
Details	Not found	50 wind turbines	9223 panels
Installed power (MW)	5.0	105.0	3.0
Assured energy (MWh/year)	31,500.0	416,678.0	14,560.0
Area utilization (km^2)	325.0		0.05
Cost of investment (R$)	12.000.000,00	500.000.000,00	20.000.000,00
Supply capability (population)	13,000.0	310,000.0	5721.0

Sustainability indicators		Unit	Açude Castanhão	Praia Formosa	Grupo Telles
Economic	Implementation cost	USD/kW	750.00	1488.10	2083.30
	Price of generated energy[g]	R$/MWh	227,00	223,00	249,00
	Payback time	Years	1.7	5.4	5.5
Technical	Occupied area	MW/km^2	0.02		60.0
	Lifetime	Years		20.0	20.0
	Generation efficiency	%	Not relevant	Not relevant	Not relevant
Environmental	GHG emissions	kg CO_2eq	400,050.0	4,945,967.9	1,368,640.0
	Emissions of other gases	kg NO_x	567.0	22,709.0	3145.0
		kg SO_2	504.0	22,083.9	3028.5
	Annual water consumption	t/year	1,134,000.0	416,678.0	145,600.0

[a]SEINFRA (2017)
[b]Prefeitura de água doce (2017)
[c]IMPSA (2017)
[d]Agência Nacional de Águas (ANA) (2010)
[e]Filippim et al. (2013)
[f]Tractebel Energia (2014)
[g]Agência Nacional de Energia Elétrica (ANEEL) (2016)
[h]Fundação Getúlio Vargas (FGV) (2009)
[i]SENAI (2016)

References

Agência Nacional de Águas (ANA) (2010) Declaração de Reserva de Disponibilidade Hídrica para o aproveitamento hidrelétrico Garibaldi, localizado no rio Canoas. Nota Técnica Conjunta ANA—DRHI-SDS-SC. ANA

Agência Nacional de Energia Elétrica (ANEEL) (2016a) Últimas Notíciais: Leilão n° 9/2015 é homologado parcialmente. Available at http://www2.aneel.gov.br/aplicacoes/noticias/Output_Noticias.cfm?Identidade=9060&id_area=90. Accessed 22 Feb 2017

Agência Nacional de Energia Elétrica (ANEEL) (2017) Últimas Notíciais: Leilão n° 9/2015 é homologado parcialmente. ANEEL, 2016. Disponível em: http://www2.aneel.gov.br/aplicacoes/noticias/Output_Noticias.cfm?Identidade=9060&id_area=90. Accessed 22 Feb 2017

Filippim ES, Mocelim JP, Fischer A, Bernardy RJ (2013) As perspectivas de atingidos por hidrelétricas: O caso da construção da Usina Garibaldi. Revista Pensamento & Realidade

Fundação Getúlio Vargas (FGV) (2009) Canal de Energia: Siif Énergies investe R$ 500 milhões na Central Eólica Praia Formosa. Centro de Estudos em Sustentabilidade. FGV-EAESP. Available at http://gvces.com.br/siif-energies-investe-r-500-milhoes-na-central-eolica-praia-formosa?locale=pt-br. Accessed 20 Feb 2017

IMPSA. Parque Eólico Santa Catarina—Brasil. Available at www.impsa.com. Accessed 9 Feb 2017

Prefeitura de Água Doce (2017) Turismo: Complexo Eólico Água Doce. Available at http://www.aguadoce.sc.gov.br/turismo/item/detalhe/2245. Accessed 10 Feb 2017

SEINFRA (2017) Pequena Central Hidrelétrica no Castanhão. Governo do Estado do Ceará. Available at http://www.seinfra.ce.gov.br/index.php/energia/pequena-central-hidreletrica-no-castanhao. Accessed 23 Feb 2017

SENAI—Departamento Regional do Ceará (2016) Notícias: Grupo Telles inaugura usina solar em Pindoretama. SENAI/CE. Available at http://www.senai-ce.org.br/99662/fiecnoticias/grupo-telles-inaugura-usina-solar-em-pindoretama. Accessed 20 Feb 2017

Tractebel Energia S. A. (2014) Comunicado ao mercado, Florianópolis

Appendix B
Calculation of AIS

J. F. G. Oliveira and T. C. G. Trindade, *Sustainability Performance Evaluation of Renewable Energy Sources: The Case of Brazil*,
https://doi.org/10.1007/978-3-319-77607-1

Santa Catarina												
Sustainability indicators			UHE	WPP	PSPP	Average	NP			Weight		
Type	Indicator	Unit					HPP	WPP	PSPP	MON	ECO	SUS
Economic	Implementation cost	USD/kW	1,547.00	1,817.00	5,865.00	3,076.00	−0.50	−0.59	−1.91	0.45	0.07	0.25
	Price of energy generated	R$/MWh	70,93	69,69	77,8	72,81	0.97	0.96	1.07	0.45	0.07	0.25
	Payback time	Years	5.75	9.38	15.53	10.22	−0.56	−0.92	−1.52	0.45	0.07	0.25
Technical	Assured energy	MWh assrd./MW	3,794	2,783	4,242	3,606	1.05	0.77	1.18	0.45	0.10	0.25
	Resource availability	Qualitative	3	3	1	2.33	1.29	1.29	0.43	0.45	0.10	0.25
	Area utilization	MW/km^2	7.18	7.5	30	14.89	0.48	0.50	2.01	0.45	0.10	0.25
	Lifetime	Years	100	20	20	46.67	2.14	0.43	0.43	0.45	0.10	0.25
Quantitative environmental	Greenhouse gases emission	kg CO_2 eq/MWh	12.7	11.87	94	39.52	−0.32	−0.30	−2.38	0.05	0.41	0.25
	SO_2 emissions	kg NO_x/MWh	0.018	0.055	0.216	0.10	−0.19	−0.57	−2.25	0.05	0.41	0.25
	NO_x emissions	kg SO_2/MWh	0.016	0.053	0.208	0.09	−0.17	−0.57	−2.25	0.05	0.41	0.25
	Annual water consumption	t/MWh	36	1	10	15.67	−2.30	−0.06	−0.64	0.05	0.41	0.25
Qualitative environmental	Biodiversity loss		5	1	1	2.33	−2.14	−0.43	−0.43	0.05	0.41	0.25
	Dismantling conservation units		5	2	1	2.67	−1.88	−0.75	−0.38	0.05	0.41	0.25

(continued)

(continued)

Santa Catarina												
Sustainability indicators			UHE	WPP	PSPP	Average	NP			Weight		
Type	Indicator	Unit					HPP	WPP	PSPP	MON	ECO	SUS
	Interaction of structure with fauna		3	2	1	2.00				0.05	0.41	0.25
Social	Toxins		2	1	3	2.00	−1.00	−0.50	−1.50	0.05	0.41	0.25
	Population removal		5	1	1	2.33	−2.14	−0.43	−0.43	0.05	0.41	0.25
	Land-use impact		5	2	1	2.67	−1.88	−0.75	−0.38	0.05	0.41	0.25
	Visual impact		1	3	1	1.67	−0.60	−1.80	−0.60	0.05	0.41	0.25
	Noise		2	3	1	2.00	−1.00	−1.50	−0.50	0.05	0.41	0.25
										AIS		

Santa Catarina											
Sustainability indicators			CN-MON			CN-ECO			CN-SUST		
Type	Indicator	Unit	HPP	WPP	PSPP	HPP	WPP	PSPP	HPP	WPP	PSPP
Economic	Implementation cost	USD/kW	−0.23	−0.27	−0.86	−0.04	−0.04	−0.13	−0.13	−0.15	−0.48
	Price of energy generated	R$/MWh	0.44	0.43	0.48	0.07	0.07	0.07	0.24	0.24	0.27
	Payback time	Years	−0.25	−0.41	−0.68	−0.04	−0.06	−0.11	−0.14	−0.23	−0.38
Technical	Assured energy	MWh assur./MW	0.47	0.35	0.53	0.11	0.08	0.12	0.26	0.19	0.29
	Resource availability	Qualitative	0.58	0.58	0.19	0.13	0.13	0.04	0.32	0.32	0.11
	Area utilization	MW/km^2	0.22	0.23	0.91	0.05	0.05	0.20	0.12	0.13	0.50
	Lifetime	Years	0.96	0.19	0.19	0.21	0.04	0.04	0.54	0.11	0.11
Quantitative environmental	Greenhouse gases emission	kg CO_2 eq/MWh	−0.02	−0.02	−0.12	−0.13	−0.12	−0.98	−0.08	−0.08	−0.59
	SO_2 emissions	kg NO_x/MWh	−0.01	−0.03	−0.11	−0.08	−0.23	−0.92	−0.05	−0.14	−0.56
	NO_x emissions	kg SO_2/MWh	−0.01	−0.03	−0.11	−0.07	−0.24	−0.92	−0.04	−0.14	−0.56
	Annual water consumption	t/MWh	−0.11	0.00	−0.03	−0.94	−0.03	−0.26	−0.57	−0.02	−0.16
Qualitative environmental	Biodiversity loss		−0.11	−0.02	−0.02	−0.88	−0.18	−0.18	−0.54	−0.11	−0.11
	Dismantling conservation units		−0.09	−0.04	−0.02	−0.77	−0.31	−0.15	−0.47	−0.19	−0.09

(continued)

(continued)

Santa Catarina											
Sustainability indicators			CN-MON			CN-ECO			CN-SUST		
Type	Indicator	Unit	HPP	WPP	PSPP	HPP	WPP	PSPP	HPP	WPP	PSPP
	Interaction of structure with fauna										−0.13
	Toxins		−0.05	−0.03	−0.08	−0.41	−0.21	−0.62	−0.25	−0.13	−0.38
Social	Population removal		−0.11	−0.02	−0.02	−0.88	−0.18	−0.18	−0.54	−0.11	−0.11
	Land-use impact		−0.09	−0.04	−0.02	−0.77	−0.31	−0.15	−0.47	−0.19	−0.09
	Visual impact		−0.03	−0.09	−0.03	−0.25	−0.74	−0.25	−0.15	−0.45	−0.15
	Noise		−0.05	−0.08	−0.03	−0.41	−0.62	−0.21	−0.25	−0.38	−0.13
			1.44	0.67	0.15	−5.71	−3.29	−4.77	−2.56	−1.56	−2.63

Ceará												
Sustainability Indicators			UHE	WPP	PSPP	Average	NP			Weight		
Type	Indicator	Unit					HPP	WPP	PSPP	MON	ECO	SUS
Economic	Implementation cost	USD/kW	750	1,488	2,083	1,440	−0.52	−1.03	−1.45	0.45	0.07	0.25
	Price of energy generated	R$/MWh	70,94	69,69	77,81	72,81	0.97	0.96	1.07	0.45	0.07	0.25
	Payback time	Years	1.68	5.38	5.52	4.19	−0.40	−1.28	−1.32	0.45	0.07	0.25
Technical	Assured energy	MWh assrd./MW	6,300	3,968	4,242	4,837	1.30	0.82	0.88	0.45	0.10	0.25
	Resource av.	Qualitative	1.00	3.00	3.00	2.33	0.43	1.29	1.29	0.45	0.10	0.25

(continued)

(continued)

Ceará												
Sustainability Indicators			UHE	WPP	PSPP	Average	NP			Weight		
Type	Indicator	Unit					HPP	WPP	PSPP	MON	ECO	SUS
	Area utilization	MW/km^2	0.02	7.50		22.51	0.00	0.33	2.67	0.45	0.10	0.25
	Lifetime	Years	32	20	20	24	1.33	0.83	0.83	0.45	0.10	0.25
Quantitative Environmental	Greenhouse gases emission	kg CO_2eq/MWh	12.7	11.87	94.00	39.52	−0.32	−0.30	−2.38	0.05	0.41	0.25
	SO_2 emissions	kg NO_x/MWh	0.02	0.05	0.22	0.10	−0.19	−0.57	−2.25	0.05	0.41	0.25
	NO_x emissions	kg SO_2/MWh	0.02	0.05	0.21	0.09	−0.17	−0.57	−2.25	0.05	0.41	0.25
	Annual water consumption	t/MWh	36	1.00	10.00	15.67	−2.30	−0.06	−0.64	0.05	0.41	0.25
Qualitative Environmental	Biodiversity loss		5	3.00	1.00	1.00	1.67	−1.80	−0.60	−0.60	0.05	0.41
	Dismantling conservation units		5	3.00	2.00	1.00	2.00	−1.50	−1.00	−0.50	0.05	0.41
	Interaction of structure with fauna		3	3.00	2.00	1.00	2.00	−1.50	−1.00	−0.50	0.05	0.41
	Toxins		2	2.00	1.00	3.00	2.00	−1.00	−0.50	−1.50	0.05	0.41
Social	Population removal		5	3.00	1.00	1.00	1.67	−1.80	−0.60	−0.60	0.05	0.41
	Land-use impact		5	3.00	2.00	1.00	2.00	−1.50	−1.00	−0.50	0.05	0.41
	Visual impact		1	750	1488	2083	1440.48	−0.52	−1.03	−1.45	0.45	0.07
	Noise		2	70.94	69.69	77.81	72.81	0.97	0.96	1.07	0.45	0.07
										AIS		

Ceará											
Sustainability Indicators			CN-MON			CN-ECO			CN-SUST		
Type	Indicator	Unit	HPP	WPP	PSPP	HPP	WPP	PSPP	HPP	WPP	PSPP
Economic	Implementation cost	USD/kW	−0.23	−0.46	−0.65	−0.04	−0.07	−0.10	−0.13	−0.26	−0.36

(continued)

(continued)

Ceará											
Sustainability Indicators			CN-MON			CN-ECO			CN-SUST		
Type	Indicator	Unit	HPP	WPP	PSPP	HPP	WPP	PSPP	HPP	WPP	PSPP
	Price of energy generated	R$/MWh	0.44	0.43	0.48	0.07	0.07	0.07	0.24	0.24	0.27
	Payback time	Years	−0.18	−0.58	−0.59	−0.03	−0.09	−0.09	−0.10	−0.32	−0.33
Technical	Assured energy	MWh assur./MW	0.59	0.37	0.39	0.13	0.08	0.09	0.33	0.21	0.22
	Resource av.	Qualitative	0.19	0.58	0.58	0.04	0.13	0.13	0.11	0.32	0.32
	Area utilization	MW/km^2	0.00	0.15	1.20	0.00	0.03	0.27	0.00	0.08	0.67
	Lifetime	Years	0.60	0.38	0.38	0.13	0.08	0.08	0.33	0.21	0.21
Quantitative Environmental	Greenhouse gases emission	kg CO_2eq/MWh	−0.02	−0.02	−0.12	−0.13	−0.12	−0.98	−0.08	−0.08	−0.59
	SO_2 emissions	kg NO_x/MWh	−0.01	−0.03	−0.11	−0.08	−0.23	−0.92	−0.05	−0.14	−0.56
	NO_x emissions	kg SO_2/MWh	−0.01	−0.03	−0.11	−0.07	−0.24	−0.92	−0.04	−0.14	−0.56
	Annual water consumption	t/MWh	−0.11	0.00	−0.03	−0.94	−0.03	−0.26	−0.57	−0.02	−0.16
Qualitative Environmental	Biodiversity loss		0.25	−0.09	−0.03	−0.03	−0.74	−0.25	−0.25	−0.45	−0.15
	Dismantling conservation units		0.25	−0.08	−0.05	−0.03	−0.62	−0.41	−0.21	−0.38	−0.25
	Interaction of structure with fauna		0.25	−0.08	−0.05	−0.03	−0.62	−0.41	−0.21	−0.38	−0.25
	Toxins		0.25	−0.05	−0.03	−0.08	−0.41	−0.21	−0.62	−0.25	−0.13
Social	Population removal		0.25	−0.09	−0.03	−0.03	−0.74	−0.25	−0.25	−0.45	−0.15
	Land-use impact		0.25	−0.08	−0.05	−0.03	−0.62	−0.41	−0.21	−0.38	−0.25
	Visual impact		0.25	−0.23	−0.46	−0.65	−0.04	−0.07	−0.10	−0.13	−0.26
	Noise		0.25	0.44	0.43	0.48	0.07	0.07	0.07	0.24	0.24
			0.72	0.39	1.15	−5.30	−3.67	−4.81	−2.64	−1.90	−2.21

Appendix C
Summary of Results' Radar Charts

See Figs. C.1, C.2, C.3, C.4, C.5 and C.6.

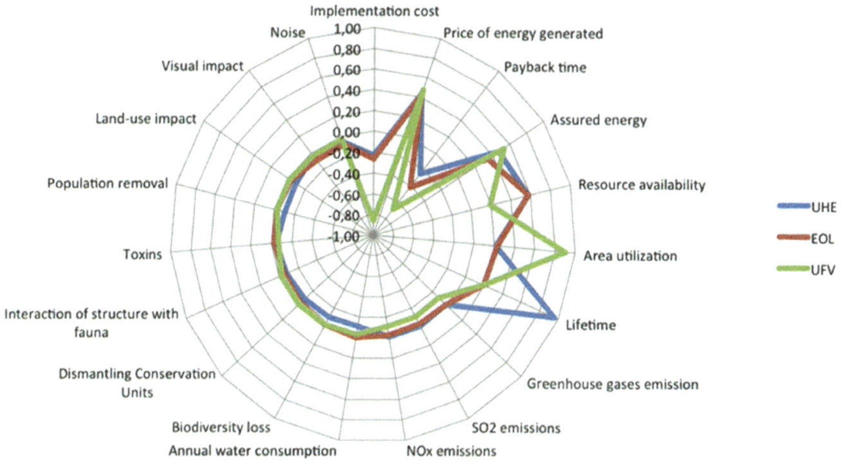

Fig. C.1 Sustainability radar chart for Scenario 1, SC. *Source* The author

J. F. G. Oliveira and T. C. G. Trindade, *Sustainability Performance Evaluation of Renewable Energy Sources: The Case of Brazil*,
https://doi.org/10.1007/978-3-319-77607-1

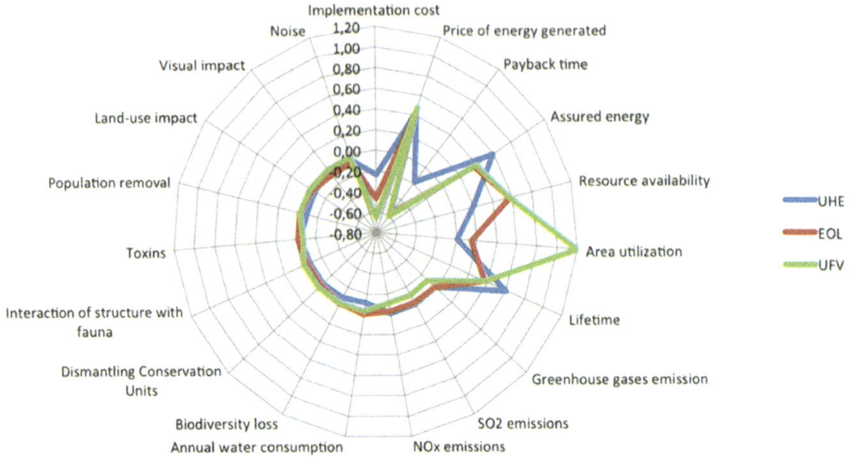

Fig. C.2 Sustainability radar chart for Scenario 1, CE. *Source* The author

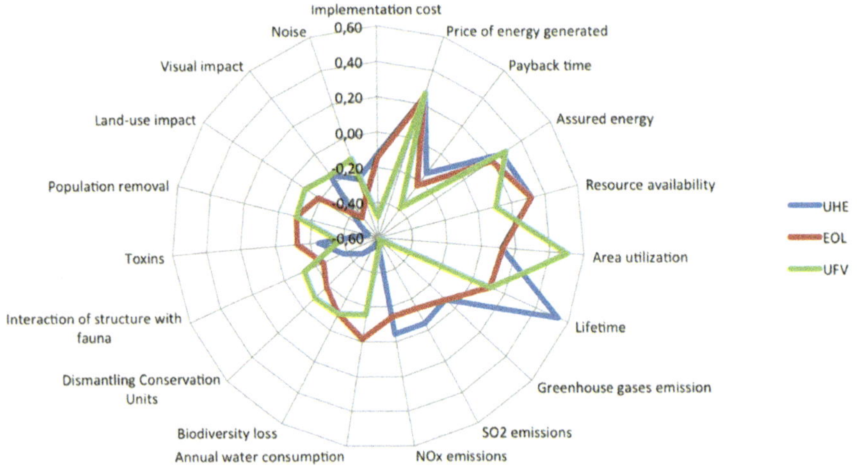

Fig. C.3 Sustainability radar chart for Scenario 2, SC. *Source* The author

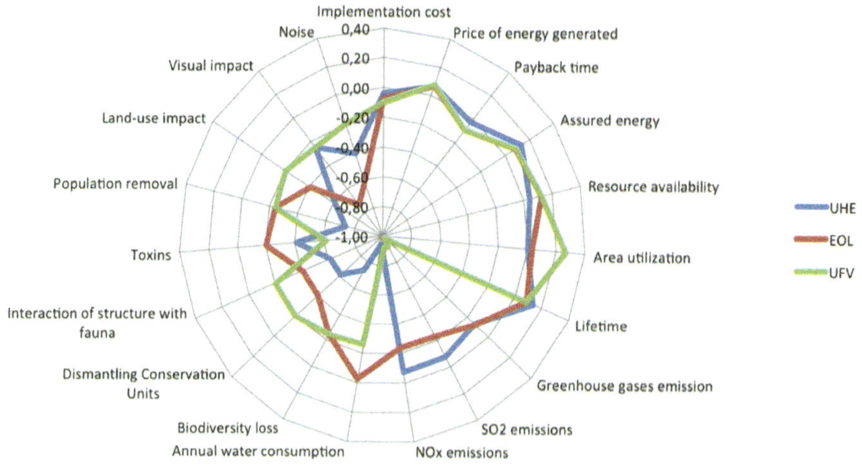

Fig. C.4 Sustainability radar chart for Scenario 2, CE. *Source* The author

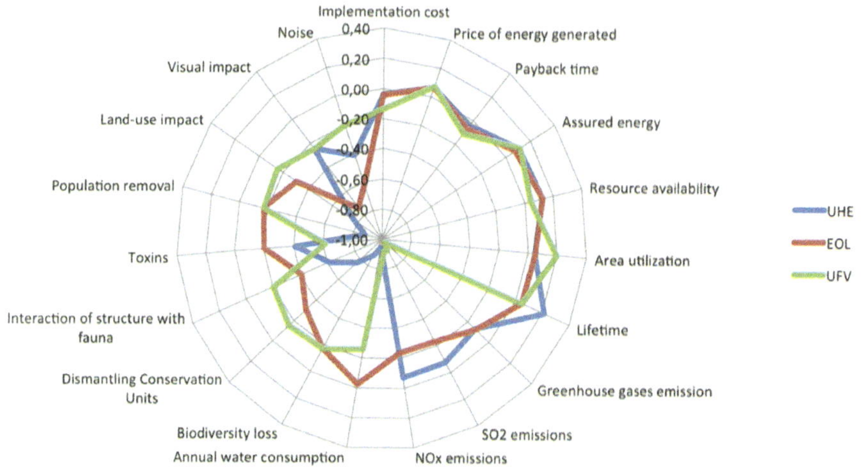

Fig. C.5 Sustainability radar chart for Scenario 3, SC. *Source* The author

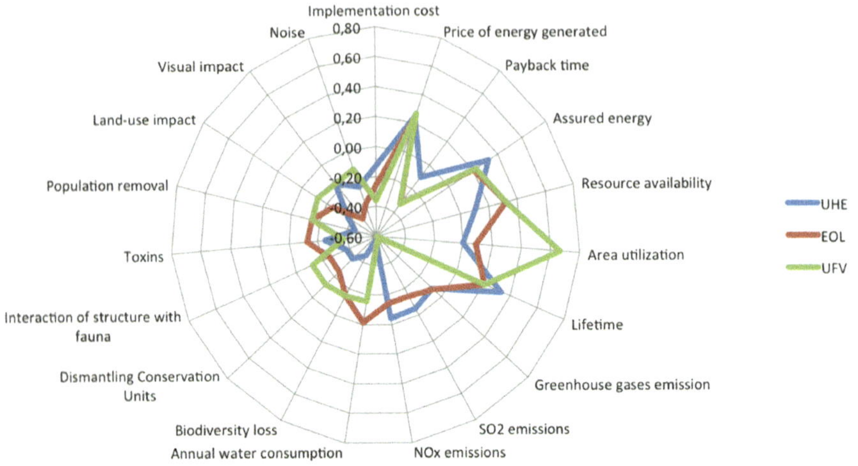

Fig. C.6 Sustainability radar chart for Scenario 3, CE. *Source* The author

The manufacturer's authorised representative in the EU is Springer Nature Customer Service Centre GmbH, Europaplatz 3, 69115 Heidelberg, Germany. If you have any concerns regarding our products, please contact ProductSafety@springernature.com

Printed and bound by CPI Group (UK) Ltd, Croydon, CR0 4YY
30/09/2026
02225664-0016